FIBER OPTICS *and*
OPTOELECTRONIC DEVICES

FIBER OPTICS *and*
OPTOELECTRONIC DEVICES

S. MOHAN, Ph.D, D.Sc.,

Former Vice–Chancellor, PRIST University, Thanjavur
Senior Professor of Materials Science
Hawassa University, Hawassa, Ethiopia
Dean-Research, Vel Tech University,
Avadi, Chennai

V. ARJUNAN, Ph.D.,

Kanchi Mamunivar Centre for Post-Graduate Studies,
Pondicherry

SUJIN P. JOSE, Ph.D.,

Department of Computational Physics, School of Physics,
Madurai Kamaraj University,
Madurai

MJP Publishers

Chennai   New Delhi   Tirunelveli

ISBN 978-81-8094-202-0

All rights reserved
Printed and bound in India

MJP 182

Publisher : J.C. Pillai

MJP Publishers

New No. 5 Muthu Kalathy Street,
Triplicane,
Chennai 600 005

© Publishers, 2014

This book has been published in good faith that the work of the author is original. All efforts have been taken to make the material error-free. However, the author and publisher disclaim responsibility for any inadvertent errors.

PREFACE

Most fascinating subject of this century is fiber optics and opto-electronics due to their elegance, simplicity, durability, low loss, improved bandwidths, high degree of security, immunity to cross talk and electrical interference, size and light weight, narrow spectral line width, high capacity and low cost. Fibre is an integral part of modern day communication. The field of fiber optics has undergone tremendous growth and advancement over the last three decades. Fiber-optic lines have revolutionized long-distance phone calls, cable TV and the internet. Broadband stands for broad bandwidth in communication path. The broader the frequency region for communication the larger is the capacity of information. The optoelectronics industry has brought about such products as compact disc players, laser printers, bar code scanners and laser pointers. The fiber optic communication industry has literally revolutionized the telecommunication industry by providing higher performance, more reliable telecommunication links with ever decreasing bandwidth cost. Fiber optics not only made a big revolution in communication but also fiber optic sensors which are widely sued in industries for measuring pressure , temperature, flow rate and strain measurement as well as in medical field. Optical fibers are also attractive for applications in sensing, control and instrumentation. An optical fiber sensing system is basically composed of a light source, optical fiber; a sensing element or transducer and a detector. Due to the development in optical communication, new types of lasers, optical amplifiers, fiber optic components and optoelectronic integrated circuits emerged. The development took place in fiber optics and opto-electronic devices are amazing.

Admiring the developments in this field, the authors of this book wanted to amalgamate the recent developments together in this book. Further, their long teaching experience in this subject made them to undertake this work. Altogether, the book consists of six chapters. The first chapter is devoted to fundamentals covering the basics of light and optical fibers. This chapter clearly explains in detail fiber characteristics, different types of fibers and losses in fibers. The second chapter deals with optical sources and detectors. It explains the importance, structure, efficiency and uses of LEDs. For continuation sake, the laser fundamentals are discussed along with various types of solid state lasers. Semiconductor lasers used as sources for long distant communication are elaborated in detail in this chapter. Optical detectors, PIN detector, Avalanche detector are discussed along with detector characteristics. Finally optical communication and connecting techniques in fibers namely splicing and connectors are also discussed in this chapter.

Chapter-III discusses optoelectronic modulators and interferometric optical sensors. Electroptic, acoutoptic, magnetoptic and fiber modulators are explained in this chapter with recent developments. Recent developments in Interferometric sensors namey Fabry-Perot, Mach Zehnder, Michelson and Sagnac interferometric sensors are explained in this chapter. For optical communication to be successful, optical amplifier and networking components are very important. Hence Chapter-IV is completely devoted to these topics Along with basics, semiconductor amplifier, rare earth doped amplifier and Raman amplifier are explained in this chapter. Added to these, communication components such as filters, attenuators, circulators, isolators, switches, couplers, splitters, wavelength converters are explained along with transmitters and receivers. Optical networking, WDM, DWDM, SONNET, SDH, AON and PON are also discussed in detail in this chapter.

Optoelectronic integrated circuits are emerging technology to boost lightwave systems and progressing very rapidly. Chapter-V gives details regarding hybrid and monolithic integration and integrated transmitters and receivers. Authors believe that this chapter will stimulate readers for further research and study on this topic. The last Chapter is completely devoted to the applications of optical fibers. Optical fibers are very familiar due to its multi- advantages. It is used in all most all branches of science starting from industry to medical field. It is impossible to discuss all the applications of optical fiber in a book of this type. However, authors concentrated on more important and common applications of fiber in industry and medical field in this chapter. Since lasers are developing very fast with wide range of wavelengths, we need sophisticated fibers for medical use and communication. Now researchers concentrate on new fibers for further development. Hence, authors added a title 'new fibers' in this chapter.

We thank our colleagues whose comments and discussion have contributed a lot for the present form of this book. We are also thankful to several learned professors, engineers and networking personal who were kind enough to offer their valuable criticisms for the improvement of this book. Many Professors at home and abroad have given support through their lecture notes and advices then and there. One of the authors (S.M) affectionately thank his grand daughter P. Mithra for showing a new life to him as well as stimulating him to take up this venture. Last but not least, our interaction with students in the last three decades and Prof. S. Mohan's service at Asian Institute of Medicine, Science and Technology University, Malaysia were also useful while preparing the manuscript of this book. Authors earnestly request readers to send their suggestions and comments to improve this book. Certainly, we will see tremendous development in fiber optic communications and sensor technology in the years to come.

Finally, We express our sincere thanks to Mr. J. C. Pillai and Mr. C. Janarathanan, MJP Publishers, Chennai for their untiring support, encouragement and interest in publishing books on frontier subjects like the present one.

S. Mohan

V. Arjunan

Sujin P.Jose

ACKNOWLEDGEMENTS

We are very grateful to learned Professors, engineers from various Universities and networking specialists for their close interaction and informative lecture notes which were useful in designing this book. We are particularly thankful to research journals and the authors of several papers which are cited under reference for their invaluable contribution to this emerging technology. Several Professors were kind enough to offer their valuable suggestions and advices for improving the manuscript. We are also thankful to the several well known trade journals, Photonics spectra and Laser focus world which provided a good amount of information on the recent developments in this field of study and it is gratefully acknowledged.

ABOUT THE AUTHORS

Mohan Sriramulu; B.Sc, M.Sc, Ph.D, D.Sc, Dip. in German, H.R.D, Buss. Management and Administration; Educationist; b April 3, 1947 Chennai, Educ. Madras Univ; Teaching Research Fellow 1969–72; Asst. Prof., Presidency College 1973–80; Reader, Anna University 1980–87; Prof., Pondicherry University 1988–2002, 2005–06; Prof., Asian Inst. Med. Sci. and Tech. (Malaysia) 2003–04; Director–Res. and Dev., PR Inst. of Sci. & Tech. 2006–2007; Vice Chancellor, PRIST University, Thanjavur 2008–2010, Dean, Senior Professor of Materials Science, Hawassa University, Ethiopia 2010– 2012, Dean-Research, VelTech University, 2013, President: Spectroscopic Soc. of India, Member: Indian Chemical Soc., New York Academy of Scis., Indian Soc. for Experiential Learning, Soc. for Chemists–France, Indian Physics Association, Soc. for the Progress of Sci., International Consortium for Experiential Learning, Laser and Spectroscopic soc. of India, Indian Council for Research in Educational Media; Patron–Spectrophysics Assn. of India, Vice President–Photonics Soc. of India; Publs–More than 800 papers published in various International and National journals of repute and 14 books; Guided 72 Ph.D scholars; organized 10 National conferences, 6 Refresher courses, one Summer school and third International Conf. on Experiential learning at Pondicherry 2002 and International Conf. on Photonics, Nanotechnology and Computer applications 2009. Contributed several papers to Experential learning; Awards–International Man of the Year 1997, 1998; Best Researcher ISPA award 1999; Scientist of the year 2001 award; Meritorious Educational Excellence award, 2002; Best scientist of the year 2004 award; Best teacher award 2004; Member in the Editorial board of Spectrochim Acta; INCONS 2005 award for Best Teacher, Researcher and Administer, *Address (Res)* 7, 3 rd *cross, Kumaran Nagar Extension,*

Lawspet, Pondicherry 605 008. (off) Dean, Research, Vel Tech University, Avadi, Chennai 600 062

Email: smoh14@rediffmail.com; smoh1947@rediffmail.com

Arjunan Velu; M.Sc, M.Phil, M.Ed, Ph.D, P.G.D.C.A, Educationist; b March 15, 1964 Thanjavur, Educ. Madras Univ & Pondicherry Univ; Associate Prof. of Chemistry 1988-till date, Govt. of Pondicherry; Publs–More than 76 papers published in various reputed International and National journals; Guided 4 Ph.D scholars; 7 Ph.D. scholars doing research under his guidance; His research areas include Materials Science, High temperature superconductors, Fiber Optics, Environmental Chemistry and Quantum Chemical Calculations on biological molecules and superconductors; *Address (Res) MIG D-26,Suthanthira Ponvizha Nagar, Pondicherry* 605 011. (off) Associate Prof. of Chemistry, Kanchi Mamunivar Centre for Post-Graduate Studies, Pondicherry 605 008.

Email: varjunftir@yahoo.com

Sujin P. Jose; M.Sc, Ph.D @ Suja Ravi Isaac, Educationist; b Kannur, Kerala, Educ. Kerala Univ., Trivandrum & ManonmaniamSundaranar Univ, Tirunelveli; Assistant Prof. of Physics; Madurai Kamaraj University, Madurai. She had 5 years of experience as Head of the Department, Department of Electronics and Communication Engineering, Vickram College of Engineering, Madurai. Publs–More than 25 papers published in various reputed International and National journals. Her research interests includesComputational Physics, Nanoscience and Biomedical Engineering. Currently she is working on nanostructured materialsand nanocompositesfor energy storage and biomedical applications and Quantum chemical investigation of polyatomic molecules. *Address (Res) Mystical Rose, 34 - Ramnagar 2nd Street, S.S. Colony, Madurai* 625 010 (off) Assistant Prof., Department of Computational Physics, School of Physics, Madurai KamarajUniversity, Madurai 625 021.

E mail: sujamku@gmail.com

ABBREVIATIONS

μm	Micrometre
ADP or $NH_4H_2PO_4$	Ammonium dihydrogen phosphate
$AgGaS_2$	Silver Gallium Sulfide
$AgGaSe_2$	Silver Gallium Selenite
Al_2O_3	Sapphire
AlGaAs	Aluminium Gallium Arsenide
AON	Active Optical Networks
APD	Avalanche photodiode
ArF	Argon fluoride
ATM	Asynchronous transfer mode
AVD	Axial vapour deposition
BBO or b–BaB_2O_4	Beta barium borate
BCBF or $BaCaBO_3F$	Barium calcium borate fluoride
BDP	Bandwidth distance product
$BeAl_2O_4$	Alexandrite
BER	Bit error rate
BLP	Bandwidth length product
BLSR	Bi-directional Line Switched Ring
CAS or $Ca_2Al_2SiO_7$	Calcium aluminosilicate
CdSe	Cadmiumselenium
Ce:LiCAF	Cerium doped lithium calcium aluminum fluoride
Ce:LiSAF	Cerium doped lithium strontium aluminum fluoride
CWDM	Conventional or Coarse Wavelength division multiplexing
DFB	Distributed feedback

DH	Double heterostructure
DKDP	Potassium di-deuterium phosphate
DPSSL	Diode pumped solid state lasers
DUT	Device under test
DWDM	Dense wavelength division multiplexing
EDFA	Erbium-doped fiber amplifier
EOM	Electroptic modulator
Er:YAG	Erbium-doped yttrium aluminium garnet
FAP or $Ca_5(PO_4)_3F$	Fluorapatite
FPI	Fabry–Perot interferometer
FTTB	Fiber-to-the-building
FTTC	Fiber-to-the-curb
FTTH	Fiber-to-the-home
FWHM	Full Width at Half Maximum
FWM	Four wave mixing
GaAs	Gallium arsenide
GaN	Gallium(III) nitride
GaP	Gallium phosphide
GGG or $Gd_3Ga_5O_{12}$	Gadolinium gallium garnet
GHz	Gigahertz
GSGG or $Gd_3Sc_2Ga_3O_{12}$	Gadolinium scandium gallium garnet
HBF	High birefringent fibers
HIC	Hybrid integrated circuit
Ho:YAG	
HOF	Oley optical fiber
Hz	Hertz
ILD	Injection laser diode
InP	Indium phosphide
InSb	Indium antimony
IR LED	Infrared Light-emitting diode
IVD	Inside vapour deposition
KBBF or $KBe_2BO_3F_2$	Potassium beryllium fluoroborate
KGW or $KGd(WO_4)_2$	Potassium gadolinium tungstate
KHz	Kilohertz

Km	Kilometre
KrCl	Krypton chloride
KrF	Krypton fluoride
KTA or $KTiOAsO_4$	Potassium titanyle arsenate
KTP or $KTiOPO_4$	Potassium titanyl phosphate
KYW or $KY(WO_4)_2$	Potassium yttrium tungstate
LAN	Local area networks
LBO or LiB_3O_5	Lithium triborate
LED	Light–emitting diode
LNB or $LiNbO_3$	Lithium niobate
LPG	Long period fiber gratings
LTA or $LiTaO_3$	Lithium tantalate
LTE	Line Terminating Equipment
MASER	Microwave amplification by stimulated emission of radiation
MCVD technique	Modified chemical vapour deposition
MEMS	Microelectromechanical systems
MHz	Megahertz
MMF	Multi mode fiber
MZI	Mach–Zehnder interferometers
NA	Numerical Aperture
$Nd:LiYF_4$ or Nd:YLF	Neodymium doped yttrium lithium fluoride
Nd:YAG	Neodimium doped yttrium aluminum garnet
$Nd:YVO_4$	Neodymium doped Yttrium Orthovanadate
NEP	Noise equivalent power
NWDW	Narrowband Wavelength division multiplexing
OCVD	Outside chemical vapour deposition
OEIC	Optoelectronic integrated circuit
OVD	Outside vapour deposition
PC	Polarization controller
PCB	Printed circuit board

PCF	Phonic crystal fiber
PDH	Plesiochronous Digital Hierarchy
PMCVD	Plasma enhanced modified chemical vapour deposition technique
PMF	Polarization maintaining fibers
PMPCF	Polarization-maintaining photonic crystal fibers
PON	Passive Optical Network
KTP	Potassium titanyl phosphate
PPLN or $LiNbO_3$	Periodically poled lithium niobate
PTE	Path Terminating Equipment
SBBO or $Sr_2Be_2B_2O_7$	Strontium beryllium borate
SBS	Stimulated Brillouin Scattering
SDH	Synchronous Digital Hierarchy
S–FAP or $Sr_5(PO_4)_3F$	Sr-fluorapatite
SI	Sagnac interferometers
SiO_2	Sicondioxide
SNR or S/N	Signal to noise ratio
SOA	Semiconductor Optical Amplifier
SONET	Synchronous Optical Networking
SPM	Self phase modulation
SPR	Surface Plasmon resonance
SRS	Stimulated Raman Scattering
SSL	Solid state laser
STE	Section Terminating Equipment
STM	Synchronous Transport Modules
$Ti:Al_2O_3$	Titanium sapphire
$U:CaF_2$	Uranium doped calcium fluoride
UPSR	Unidirectional Path Switched Ring
UV LED	Ultraviolet light emitting diode
VCSEL	Vertical-cavity surface-emitting laser
VLSI	Very Large Scale Integration
VSWR	Voltage standing wave ratio
WDM	Wavelength division multiplexing

XeCl	Xenon chloride
XeF	Xenon fluoride
XPM	Cross phase modulation
YAG or $Y_3Al_5O_{12}$	Yttrium aluminum garnet
Yb	Ytterbium
Yb:GdCOB	Ytterbium dopped gadolinium calcium oxyborate.
YCOB or $YCa_4O(BO_3)_3$	Yttrium calcium oxyborate
$ZnGeP_2$	Zinc-Germanium Diphosphide

CONTENTS

Chapter–1

BASICS OF LIGHT AND FIBER

1.1 Electromagnetic Wave

Electromagnetic spectrum ranges from gamma ray, X-ray, ultraviolet, visible, infrared, microwaves and radiofrequency. Electromagnetic wave is made of discrete packets of energy called photons. Photons have no mass but carry momentum and travel at the speed of light. They have the same kind of wavy nature that repeats itself over a distance called the wavelength. Electomagnetic waves have both particle-like and wave-like properties. In the 1860's and 1870's, a Scottish scientist James Clerk Maxwell developed a theory to explain electromagnetic waves. According to him, electrical fields and magnetic fields can couple together to form electromagnetic waves. This made him to propose Maxwell's Equations.

Electomagnetic energy can be described by frequency, wavelength or energy and they are related mathematically ($E = h\nu = hc/\lambda$). Radio and microwaves are usually described in terms of frequency (Hertz), infrared and visible light in terms of wavelength (meters) and X-rays and gamma rays in terms of energy (electron volts). An electron volt is the amount of kinetic energy needed to move an electron through one volt potential. This is a scientific convention.

The number of crests that pass at a given point within one second is known as the frequency of the wave and the unit of frequency is Hertz (Hz), named after Heinrich Hertz who established the existence of radio waves. The electromagnetic waves have crests and troughs. The distance between two consecutive crests or troughs is the wavelength. The shortest wavelengths are just fractions of the size of an atom, while the longest wavelengths scientists currently study can be larger than the diameter of our planet. In electromagnetic spectrum, energy increases as the wavelength shortens ($E = h\nu = hc/\lambda$).

1.2 Nature of Light

According to classical view, a particle is a concentration of energy in space and time whereas a wave is spread out over a larger region in space and time. Newton's corpuscular theory of light explains the straight-line behaviour of sharp shadows of objects placed in the path of light beam whereas the wave theory explains interference and diffraction. Augustin Jean Fresnel showed that light is a transverse wave nature. The wave theory of light explains another important physical property of light namely polarization and it is a measurement of the electromagnetic field's alignment. The sunglasses are able to eliminate glare by absorbing the polarized portion of the light. Light travels at high speed (300,000 Km/sec) and hence information can be conveyed very quickly. The shorter the wavelength, greater the energy of a photon.

In fiber optics, the function of light is the collection, conveyance and conversion. When all these are combined together, we get communication. In fiber optic communication, low power laser is used in pulsed form. The pulse group consists of the speech as input to the encoder. Then the light is conveyed through optical fiber to the receiver for reconversion at the decoder. Of course there are limitations of light such as absorption, scattering and dispersion which we will see later in this chapter.

1.3 OPTICAL LAWS

1.3.1 Reflection

Objects are viewed from the light when they reflect. Reflected light obeys the law of reflection, namely the angle of reflection equals the angle of incidence ($q_i = q_r$). When the surface irregularities of an object are larger than the wavelength of light, then the light reflects in all directions. The reflection of light is shown in Figure 1.1.

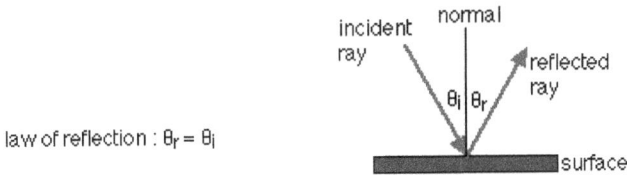

law of reflection : $\theta_r = \theta_i$

Figure 1.1 Reflection of light

1.3.2 Refraction

As light ray passes from one medium to another medium, it changes direction which is known as refraction of light. The change in direction of the light ray depends on the refractive index of the medium. Further, when light travels from one medium to another, the speed of light changes similar to the wavelength. The index of refraction (n) is expressed in terms of wavelength as follows:

$$n = \lambda / \lambda_{med}$$

where, λ is the wavelength in vacuum and λ_{med} is the wavelength in the medium. It can also be given as the ratio of the speed of light in vacuum (C) to the speed of light in the medium (C_{med}).

$$n = C / C_{med}$$

It is interesting to note that the frequency of the light is constant even if the speed and wavelength changes. The frequency (ν), wavelength (l) and velocity (c) of light are related through the equation, $\nu = c/l$

When light is travelling from medium 1 to medium 2, the angles are measured from the normal to the interface. The relation between the refracted angles of light in the second medium to the angle of incidence is given by Snell's law. According to Snell's law, the ratio of the sines of the angles of incidence and refraction is equivalent to the ratio of phase velocities in the two media or equivalent to the opposite ratio of the indices of refraction

$$\text{Sin } \theta_1 / \text{Sin } \theta_2 = \varsigma_1 / \varsigma_2 = n_2 / n_1$$

with each θ as the angle measured from the normal, ς is the velocity of light in the respective medium and n as the refractive index of the respective medium.

Further, when $\theta_1 = 0^0$ (means the ray perpendicular to the interface), $\theta_2 = 0^0$ irrespective of the values of n_1 and n_2. That is, a ray entering a medium perpendicular to the surface is never bent. The refraction of light is illustrated in Figure 1.2.

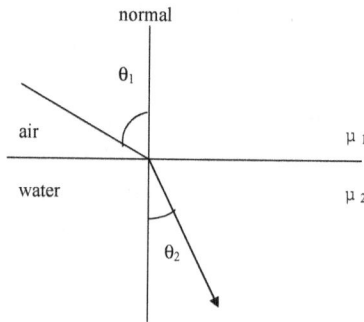

Figure 1.2 Refraction of light

1.3.3 Total Internal Reflection and Critical Angle

When light travels from denser medium to rarer medium, it undergoes refraction. To be precise, when the light crosses an interface into a medium with a higher refractive index, it bends towards the normal. Conversely, if it crosses an interface from denser medium to rarer medium, it will bend away from the normal. If the sine of the angle of refraction be greater than one, the light is completely reflected by the boundary. This phenomenon is known as total internal reflection. The largest possible angle of incidence for which the refracted ray traces the boundary is known as critical angle (c). The refractive index and the critical angle are related through the equation n = 1/Sin(c). The total internal refelection is shown in Figure 1.3.

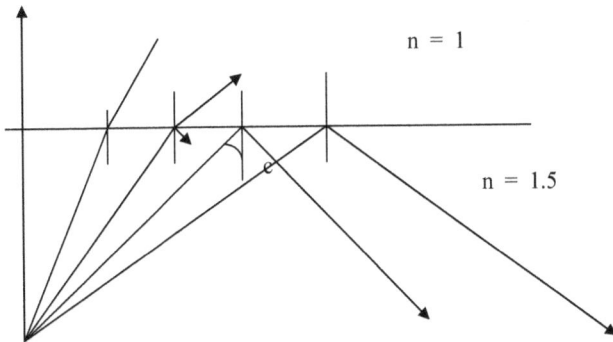

Figure 1.3   Total internal reflection

1.3.4 Dispersion

The splitting of a ray into its component colours is known as dispersion of light and the band of colours is known as a spectrum. Silica glass has refractive index which is wavelength sensitive. That is why a prism can spread white light into its constituent colours. The dispersion of white light is illustrated in Figure 1.4.

White light consists of photons of various colours or wavelengths. The wavelength of red photon is the longest and the wavelength of the violet photon is the shortest. Thus, when a white light passes through a glass medium like a prism, different photons cross the medium at different speeds. The red colour appears at the top of the spectrum because it is deviated the least or it is refracted the least (less bending). On the other hand, the violet end of the spectrum is bent the most or refracted most, as it takes longer to traverse the glass medium. In optical instruments, dispersion leads to chromatic aberration.

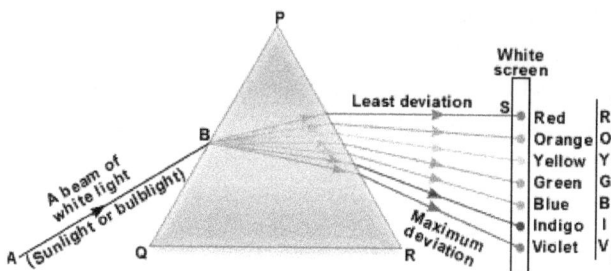

Figure 1.4   Dispersion of white light

1.3.5 Rayleigh Scattering

When the wavelength of the incident light and scattered light is same, it is known as Rayleigh scattering. It is commonly seen in the scattering of light in molecules in air. The scattering power is inversely proportional to the fourth power of the wavelength of the light ($1/\lambda^4$). Rayleigh scattering can be considered to be elastic scattering since the photon energies of the scattered photons is not changed.

The blue color of the sky is due to Rayleigh scattering when sunlight scatters the molecules of the atmosphere. Rayleigh scattering is predominant at short wavelength, viz., the blue end of the visible spectrum. The scattering from molecules and very tiny particles lesser than 1/10 of incident wavelength is seen in Rayleigh scattering. Since blue line scatters more, we never use blue light for signaling. We prefer red light for signaling as it scatters least.

1.3.6 Mie Scattering

For particle sizes larger than the wavelength, Mie scattering occurs. This scattering produces a pattern like an antenna lobe, with a sharper and more intense forward lobe for larger particles. Mie scattering is not strongly wavelength dependent and produces the almost white glare around the sun when a lot of particulate material is present in the air. It also gives us the white light from mist and fog.

1.3.7 Raman Scattering

When the incident photons interact with the molecules, the energy is either gained or lost. Hence, the scattered photons are changed in frequency which is an inelastic scattering. This type of scattering is known as Raman scattering. Like Rayleigh scattering, the Raman scattering depends upon the polarizability of the molecules. There are two types of Raman scattering, stokes and anti-stokes scattering. When molecules absorb energy, the scattered photon of lower energy generates stokes lines on the red side of the incident spectrum. On the other hand, if molecules loose energy, the incident photons are shifted to the blue side of the spectrum to form anti-stokes lines. Due to stokes and anti-stokes lines, Raman spectrum is symmetric to the Rayleigh line.

1.4 INTRODUCTION TO FIBER OPTICS

Fiber optics is a technology that uses glass or plastic hair thin strands or fibers to transmit signals through light. Fiber optics has several advantages over traditional metal wire communications lines. An optical fiber is a flexible, transparent fiber made of a pure glass (silica) not much wider than a human hair. It functions as a waveguide, or light pipe, to transmit light between the two ends of the fiber

It all began about 42 years ago in the R&D laboratories at Corning, Bell Labs and I.T.T U.K, which first installed fiber optics in Chicago, Illinois, U.S.A in 1976. The use of fiber optics was not available until 1970. Corning Glass Works was able to produce a fiber with a loss of 20 dB/Km. By the early 1980s, fiber networks connected the major cities on east coast at U.S.A. By the mid-80s, fiber was replacing all the copper, microwave and satellite links. In the present century, fiber optics is either the dominant medium or a logical choice for every communication system.

Fiber optics has several advantages over traditional communications. Fiber optic cables have a much greater bandwidth (hence carry more data) than metal cables. Fiber optic cables are less susceptible than metal cables to electromagnetic interference. Fiber optic cables are much thinner and lighter than metal wires. Here, data can be transmitted digitally rather than analogically. Summarisingly, the advantages of optical fibers are

1. Immunity to cross talk and electrical interference.

2. No cross talk.

3. Glass fibers are insulators.

4. It gives improved bandwidths. That is it carries huge amount of information.

5. They offer high degree of security.

6. They offer low losses.

7. Their small size and light weight is comfortable to use.

8. Only one fiber is required for communication.

9. Very high concentration of optical power and very little spread of that power.

10. Small antennas.

11. Coherence property of laser light.

12. Narrow spectral line width.

13. Low cost per channel.

14. Very high capacity.

The main disadvantage of fiber optics is that the cables are expensive to install. In addition, they are more fragile than wire and are difficult to splice.

Fiber optics is a popular technology for local-area networks. In future, almost all communications will be done through fiber optics. A fiber optic cable consists of a bundle of glass threads, each of which is capable of transmitting signals as light waves. Specially designed fibers are used for a variety of other applications, including sensors and fiber lasers. The infrared windows viz., 850 nm for medium distance data, 1300 nm and 1550 nm windows for long distance telecommunications are used in fiber optics.

1.5 NEED FOR FIBER OPTICS

Due to rapid growth in civilization, communication was developed at faster rate. To meet this end, copper cable communication failed due to two main reasons. One is the requirement of bandwidth for world wide web services, internet services, telephone calls, satellite and cable TVs etc., The second was due to the higher price for copper metals.

Hence, as an alternative fiber optic cables emerged with advantages over copper. They are

1. Fiber optic networks operate at high speeds up to the gigabits.
2. It has a large information carrying capacity due to bandwidth.
3. Signals can be transmitted to a considerable distance without strengthening.
4. It offers greater resistance to electromagnetic noise such as radios, motors, lightening and thunder or other nearby cables.
5. Fiber optic cables costs much less to maintain.
6. Some 10 billion digital bits can be transmitted per second along an optical fiber link in a commercial network which is enough to carry tens of thousands of telephone calls.

With the developments in research in optical fibers along with new lasers and diodes, one can expect that the commercial optical networks will carry trillions of bits of data per second in years to come. Fiber optics is steadily replacing copper wires for communication signal transmission. Basically, fiber optics use light pulses to transmit information in fiber lines instead of electronic pulses in copper lines. Local area networks (LAN) is a collective group of computers or computer systems connected to each other allowing for shared program software or data bases. Universities, Colleges, Office buildings and industrial plants make use of optical fiber within their LAN systems.

1.6 PRINCIPLE OF LIGHT PROPAGATION THROUGH OPTICAL FIBER

The transmission of light along an optical fiber can be understood from geometrical and physical optics. In geometrical optics, the light is considered as a simple ray and it gives a clear picture of the propagation of light along a fiber. It can be used to approximate the light acceptance and guiding properties of optical fibers. The electromagnetic wave (wave representation) also known as the mode theory explains the behavior of light within an optical fiber, and it is useful in describing the optical fiber properties of absorption, attenuation, and dispersion

The principle of total internal reflection is the key to understand the propagation of light in optical fiber. Optical fiber is long, thin strand of very pure glass about the diameter of human hair. Optical fibers are arranged in bundles called optical cables, and it is used to transmit light signals over long distances. It is interesting to note the basic structure of optical fibers typically include a transparent core surrounded by a transparent cladding material with a lower refractive index and buffer coating. Buffer coating provides mechanical protection and bending flexibility for the fiber. The core is the inner part of the fiber which guides light. Light is kept in the core by total internal reflection which causes the fiber to act as a waveguide.

The cladding surrounds the core completely. The refractive index of the core is higher than that of cladding so that the light in the core that strikes the boundary with the cladding at an angle shallower than critical angle. Hence, it is reflected back into the core by total internal reflection. Total internal reflection in the optical fiber is shown in Figure 1.5.

The common core diameter ranges from 8–62.5 μm for single or multimode fibers while the cladding diameter ranges from 250–900 μm. The light waves in distinct patterns in the optical fiber are known as modes. Mode describes the distribution of energy across the fiber. The exact pattern depends on the transmitted light wavelength and on the variation of refractive index in the core.

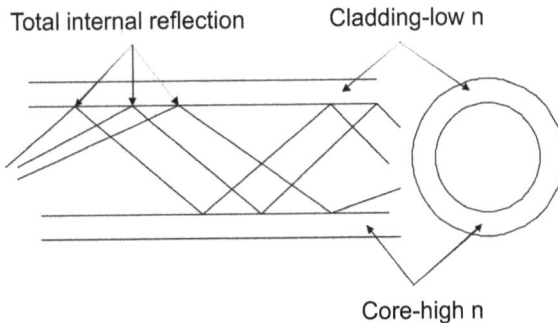

Figure 1.5 Total internal reflection in the optical fiber

1.7 TYPES OF FIBERS

1.7.1 Basic Structure of an Optical Fiber

The basic structure of an optical fiber consists of three parts; the core, the cladding, and the coating or buffer. The core is a cylindrical rod of dielectric material. Light propagates mainly along the core of the fiber. The core is generally made of glass. The core is surrounded by a layer of material called the cladding. The cladding layer is made of a dielectric material. The index of refraction of the cladding material is less than that of the core material. The cladding is generally made of glass or plastic. The cladding reduces loss of light from the core into the surrounding air. It also reduces scattering loss at the surface of the core. Further, the cladding protects the fiber from absorbing surface contaminants and adds mechanical strength. For extra protection, the cladding is enclosed in an additional layer called the coating or buffer. The coating or buffer is a layer of material used to protect an optical fiber from physical damage. The material used for a buffer is a type of plastic and it prevents abrasions. Buffer further prevents the optical fiber from scattering losses caused by microbends.

Single mode, multimode and plastic optical fibers are three types of optical fibers. Glass optical fibers are almost made from pure silica but other materials such as fluorozirconate, fluoroaluminate, germanate and chalcogenide glasses are also used for long wavelength infrared applications. In general, core must be very clear and pure material for the light such as 850, 1300 and 1500 nm.

In optical fibers, modes of fiber are characterized by numbers. Single mode fiber carries only the lowest mode and it is termed as '0', while multimode fibers carry higher order modes. The number of modes that can propagate in a fiber depends on the fiber's numerical aperture or acceptance angle as well as on its core diameter and the wavelength of light.

Numerical Aperture (NA)

The cone of acceptance is the angle within which the light is accepted into the core and is able to travel along the fiber. The half-angle of this cone is known as the acceptance angle, qmax. Calculating the cone of acceptance is not an easy task. Hence, we use the property of the fiber called numerical aperture. The numerical aperture of a fiber is a figure which represents its light gathering capability. For step-index multimode fiber, the acceptance angle is determined only by the indices of refraction.

$$NA = n \, Sin \, \theta_{max} = (n_c - n_d)^{1/2}$$

where, n is the refractive index of the medium before light enters the fiber, n_c is the refractive index of the fiber core and n_d is the refractive index of the cladding.

Therefore, the acceptance angle = $sin^{-1}(NA)$

For a step-index multimode fiber, the number modes N_m is approximated as

$$N_m = 0.5 \, [(\pi D \times NA) / \lambda \,]^{1/2}$$

Where D is the core diameter, λ is the operating wavelength and NA is the numerical aperture or acceptance angle.

The above formula is not suitable for fibers carrying only a few modes. It is a common belief that using a low numerical aperture, fiber will focus the light from a source which is not true. A narrow numerical aperture (NA) fiber simply admits less light than a wider NA fiber, assuming the source is emitting light at a wide numerical aperture.

Mode means methods of transmission. The number of modes is always a whole number. The number of modes is given by

$$\text{Number of modes} = \{\text{Diameter of core} \times NA \times \pi/\lambda\}^2 / 2$$

where, NA is numerical aperture of the fiber and λ is the wavelength of the source. If the wavelength of the transmitted light is decreased, the number of modes would increase.

For a step-index single mode fiber, the number modes N_m is approximated as

$$N_m = 0.5 \, [(\pi D / \lambda)^2 \, (n_c{}^2 - n_d{}^2)]$$

Where D is the core diameter, $- \lambda$ is the operating wavelength and n_{cl} is the refractive index of the cladding and n_c is the refractive index of the fiber core.

By reducing the core diameter sufficiently, one can limit transmission to a single mode. The following formula gives the maximum core diameter D, which limits transmission to a single mode at a particular λ.

$$D < 2.4 \, \lambda / \pi \, (n_i - n_c)^{1/2}$$

If the core is any larger, the fiber can carry two modes.

The acceptance angle in a fiber is shown in Figure 1.6.

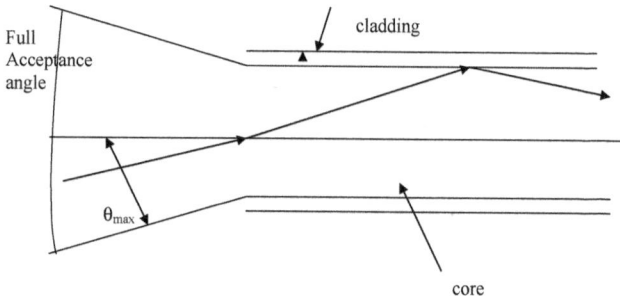

Figure 1.6   Acceptance angle in a fiber

1.7.2   Single Mode Fiber

Single mode fiber possess smaller core than multimode. Single mode fiber is a single strand of glass fiber with a diameter of 8.3 to 10 microns which propagate the lowest order mode with the typical wavelength 1300 to 1320 nm. The basic requirement for a single mode fiber is that the core should be small enough to restrict transmission to a single mode. Single mode fiber has a narrow diameter through which only one mode can propagate with wavelength 1310 or 1550 nm. In single mode fiber, the refractive index between the core and the cladding changes less than it does for multimode fiber. Light travels parallel to the axis creating very little pulse dispersion. It carries higher bandwidth than multimode fiber. Single mode fiber is also known as mono mode optical fiber, single mode optical waveguide and uni-mode fiber. Further, single mode fiber has higher transmission rate than multimode. The small core and single light wave virtually eliminates any distortion which arises due to overlapping light pulse. It provides the least signal attenuation and the highest transmission speed. Telephone and cable networks install millions of kilometers of this fiber every year. The structure of the single mode fiber is shown in Figure 1.7

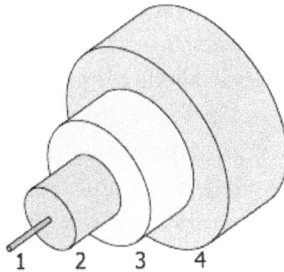

Figure 1.7 Structure of a typical single-mode fiber (1) Core (8 μm diameter), (2) Cladding (125 μm dia), (3) Buffer (250 μm dia) and (4) Jacket: (400 μm dia)

1.7.3 Advantages and Disadvantages of Single Mode Fiber

Advantages

1. Single mode fiber does not face modal dispersion, modal noise and other effects that arise with multimode transmission.

2. Single mode fiber cab carries signals at much higher speeds than multimode fibers. They are the standard choice for high data rates or long distance telecommunication (> a couple of Kms) which use diodes as source.

Disadvantages

1. Due to the smaller size of the core in single mode fiber, coupling light requires much tolerances than the coupling light in multimode fiber. Recent developments indicate that these tighter tolerances can be achievable.

2. Single mode fiber components and equipments are very expensive and hence multimode counter parts are widely used.

1.8 Multimode Fiber

Fibers that carry more than one mode are called multimode fibers. Multimode fiber has a bigger diameter. The range is around

50 to 100 micron but 62.5 micrometer is usually used. In most of the applications, two fibers are used in multimode fibers and Wavelength Division Multiplexing is not normally used on multimode fibers. Plastic optic fibers promise performances similar to glass fibers at shorter distances with lower cost. Multimode fibers give a high bandwidth at high speeds (10 to 100 Mbs, 275 m to 2 Km) over medium distances. Typical multimode fiber core diameters are 50, 62.5 and 100 micrometers. Light waves are dispersed into numerous paths or modes as they travel through

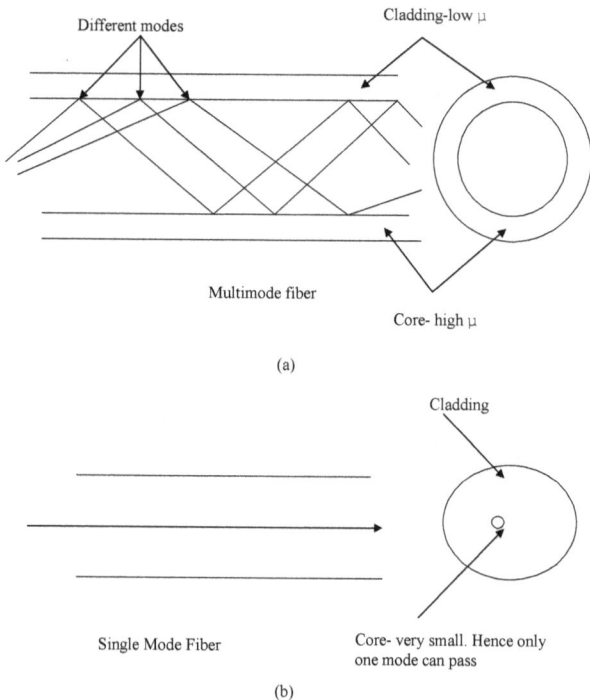

Different modes

Cladding-low μ

Multimode fiber

Core- high μ

(a)

Cladding

Single Mode Fiber

Core- very small. Hence only one mode can pass

(b)

Figure 1.8 Modes in (a) multimode and (b) single mode fiber

the fiber's core typically 850 or 1300 nm. In the long distance (>914.4 m), multiple paths of light cause signal distortion at the receiving end leading to an unclear and incomplete data transmission. Hence, single mode fibers are recommended for this transmission using gigabit and beyond. Yellow jacket represents the colour code for single mode fiber while orange jacket represents multimode fiber. The Figure 1.8 illustrates the modes in multimode and single mode fiber.

There are two types of multimode fibers. One type is step-index multimode fiber and the other is graded-index multimode fiber.

1.8.1 Step-index Multimode Fiber

Step-index fiber has a large core up to 100 microns in diameter. Hence, some of the light rays with digital pulse may travel a direct route whereas others zigzag as they bounce off the cladding. These alternative pathways cause the different groupings of light rays referred to as modes to arrive separately at a receiving point. The pulse with different modes begins to spread from its well defined shape. Therefore it is necessary to leave space between pulses to avoid overlapping. Step-index multimode fiber is best suited for transmission over short distances such as an endoscope. Step-index multimode fibers are mostly used for imaging and illumination.

1.8.2 Graded-index Multimode Fiber

Graded-index multimode fiber contains a core in which the refractive index diminishes gradually from the central axis towards the cladding. The higher refractive index at the center makes the light rays to move along the axis more slowly than those near the cladding. Further, light in the core curves helically due to the graded-index and reduce its travel distance. The shortened path and the higher speed allow light at the periphery to arrive at a receiver at about the same time as the slow but straight rays in the core axis. Hence, a digital pulse suffers less dispersion. Graded-index multimode fibers are used for data communications and networks carrying signals moderate distances-typically not more than a couple of kilometers.

1.8.3 Optical Fiber Index Profile

Optical fiber index profile is the refractive index distribution across the core and the cladding of a fiber. In step-index single mode fiber, the core has uniform refractive index while cladding has a lower refractive index. The fiber cross section and fiber refractive index profile for step-index single mode fiber is shown below (Figure 1.9). Other optical fiber has a graded-index profile, in which refractive index varies gradually as a function of radial distance from the fiber center. Graded-index profiles include power-law index profiles and parabolic index profiles. The following figure shows some common types of index profiles for single mode (Figure 1.9) and multi mode fibers (Figure 1.10).

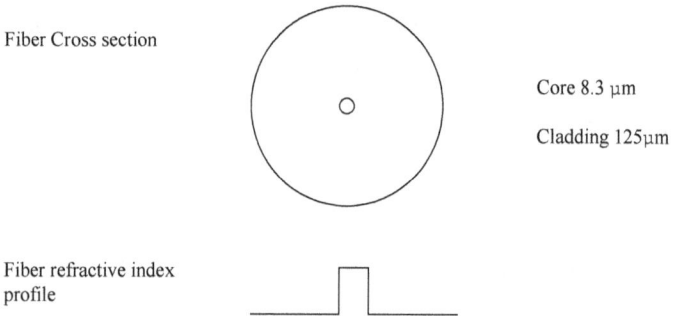

Fiber Cross section

Core 8.3 μm

Cladding 125μm

Fiber refractive index profile

Figure 1.9   Refractive index profile for step-index single mode fiber

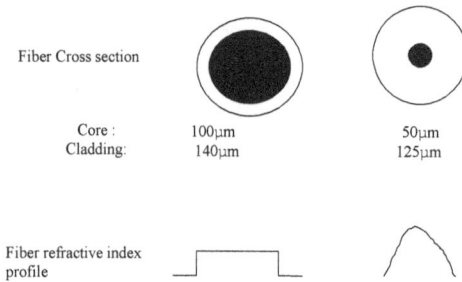

Fiber Cross section

Core :
Cladding:

100μm   50μm
140μm   125μm

Fiber refractive index profile

Figure 1.10 Refractive index profile for step-index multimode fibers.

1.9 ATTENUATION OR LOSSES IN OPTICAL FIBERS

Attenuation is the loss of optical power as light travels through fiber. The extremely low attenuation or transmission loss of optical fibers is one of the most important factors for a medium of transmission. Signal transmission within optical fibers should be free from attenuation. Optical fiber attenuation is the measurement of light loss between input and output. Total attenuation is the sum of all losses.

The signal attenuation or fiber losses are defined as the ratio of the input (transmitted) optical power into the fiber to the output (received) optical power from the fiber. The attenuation of the signal is measured in decibel/km

Attenuation in dB = $10 \log_{10} P_2 / P_1$ where, P_1 and P_2 are two divergent power levels.

Attenuation ranges over 300 dB/km for plastic fibers to around 0.21 dB/km for single mode fiber. Attenuation varies with the wavelength of light. It is due to two main causes namely absorption and scattering.

There are different types of attenuation or losses in fiber optics. They are

1. Absorption loss
2. Material or Rayleigh Scattering loss
3. Chromatic or Wavelength, Dispersion loss
4. Radiation losses
5. Modal dispersion
6. Coupling losses

Any contamination causes a power loss. The laser light shining through a fiber optic cable is subjected to loss of strength primarily through dispersion and scattering of light within the cable itself. When the laser fluctuates faster, there is a greater risk of dispersion. Repeaters are necessary to strengthen the signal in certain applications.

1.9.1 Absorption Loss

Absorption is the process by which impurities in the fiber after manufacture, absorb optical energy and dissipate it as other wavelengths or in the form of mechanical vibration (a small amount of heat). The amount converted to heat is very minor. This absorption energy is removed from the propagating pulse and given up later in some other form. The hydroxyl ions and trace elements are the worst absorbers. Extrinsic absorption results from the presence of impurities. Transition-metal impurities such as Fe, Cu, Co, Ni, Mn, V and Cr absorb strongly in the wavelength range 0.6~1.6 µm. To avoid absorption loss, it is aimed to have 1 part in 10^9 for water and 1 part in 10^{10} for metallic traces in the fiber. In new kind of glass fiber, known as *dry fiber*, the OH ion concentration is reduced to such low levels that the 1.39 µm peak almost disappears.

Light-emitting diodes (LEDs) emit light containing many wavelengths. Each wavelength within the composite light signal travels at a different velocity when propagating through glass. Consequently, light rays that are simultaneously emitted from an LED and propagated down an optical fiber do not arrive at the far end of the fiber at the same time, resulting in an impairment called chromatic distortion (sometimes called wavelength dispersion).

Chromatic distortion can be eliminated by using a monochromatic light source, such as an injection laser diode (ILD). Chromatic distortion occurs only in fibers with a single mode of transmission

1.9.2 Scattering Loss

Scattering the common source of attenuation in optical fibers is the loss of optical energy due to molecular imperfections or lack of optical purity in the fiber due to manufacturing process and from the basic structure of the fiber. Rayleigh scattering is due to small localized changes in the refractive index of the core and the cladding material. In commercial fibers operating between 700 nm and 1600 nm wavelength, the main source of loss is called Rayleigh scattering. Rayleigh scattering is the main loss mechanism between the ultraviolet and infrared regions. Loss caused by Rayleigh scattering is proportional to the fourth power of the wavelength ($1/\lambda^4$). As the wavelength increases, the loss caused by Rayleigh scattering decreases. Rayleigh scatter is most severe in light with short wavelength. If the size of the defect is greater than one-tenth of the wavelength of light, the scattering mechanism is called Mie scattering. Mie scattering, caused by these large defects in the fiber core, scatters light out of the fiber core. However, in commercial fibers, the effects of Mie scattering are insignificant. Optical fibers are manufactured with very few large defects.

Scattering scatters the light in all directions including back to the optical source. This light reflected back is what allows optical time domain reflectometers (OTDRs) to measure attenuation levels and optical breaks.

1.9.3 Microbend and Macrobend Loss

Bending the fiber also causes attenuation. Bending losses are caused by microbends and macrobends. A sharp bend in a fiber can cause significant losses as well as the possibility of mechanical failure. The bend in the short length of the fiber causes higher losses. Microbends are small microscopic bends of the fiber axis that occur mainly when a fiber is cabled. Microbend loss results from small variations or bumps in the core-to-cladding interface. Transmission losses increase due to the fiber radius decreasing to the point where light rays begin to pass through the cladding boundary. This causes the fiber rays to reflect at a different angle, therefore creating a circumstance where higher order modes are refracted into the cladding to escape. As the radius decreases, the attenuation increases.

Fibers with a graded-index profile are less sensitive to microbending than step-index types. Fibers with larger cores and different wavelengths can exhibit different attenuation values. Fiber loss caused by microbending can still occur even if the fiber is cabled correctly. During installation, if fibers are bent too sharply, macrobend losses will occur.

Microbend losses are caused by small discontinuities or imperfections in the fiber. Uneven coating applications and improper cabling procedures increase microbend loss. External forces are also a source of microbends. An external force deforms the cabled jacket surrounding the fiber but causes only a small bend in the fiber. Microbends change the path that propagating modes. Microbend loss increases attenuation because low-order modes become coupled with high-order modes that are naturally lossy (causing appreciable loss).

Macrobends are bends having a large radius of curvature relative to the fiber diameter. Macrobend losses are caused by deviations of the core as measured from the axis of the fiber. These irregularities are caused during the manufacturing procedures and should not be confused with microbends.

Fiber sensitivity to bending losses can be reduced. If the refractive index of the core is increased, then fiber sensitivity decreases. Sensitivity also decreases as the diameter of the overall fiber increases. However, increases in the fiber core diameter increase fiber sensitivity. Fibers with larger core size propagate more modes. These additional modes tend to be more lossy.

There are few uses of bending losses which are based on either the increase in the attenuation or on making use of the light which escapes from the optic fiber. A fiber optic pressure sensor is a good example.

1.9.4 Intermodal and Intramodal Dispersion

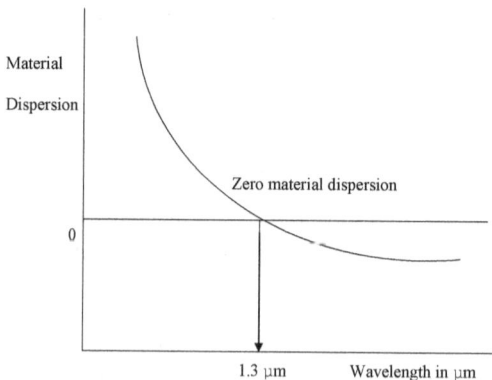

Figure 1.11   Material dispersion-wavelength in silica glass

The plot connecting material dispersion verses wavelength in silica glass is shown in Figure 1.11.

It is seen from the plot that the material dispersion effect passes through zero at about 1.3 μm. At this wavelength, optical attenuation is quite low (0.5 dB/Km). Hence, 1.3 μm is a favourable wavelength for optical fiber communication.

Modal dispersion is caused by the difference in the propagation times of light rays that take different paths down a fiber. Obviously, modal dispersion can occur only in multimode fibers. It can be reduced by using graded-index fibers and almost entirely eliminated by using single-mode step-index fibers. The design of graded-index fiber eliminates about 99% of intermodal dispersion. Here, the ray which arrives late has taken a longer route. This can be compensated by making the ray that takes longer route to move faster. If the speed and distance of each route is carefully balanced then all the rays can be made to arrive at the same time and hence no dispersion.

In a step-index multimode fiber, rays of light enter the fiber with different angles to the fiber axis, up to the fiber's acceptance angle (numerical aperture). Rays that enter with a shallower angle travel by a more direct path and arrive sooner than those that enter at steeper angles (which reflect many more times off the core/cladding boundaries as they travel the length of the fiber). The arrival of different modes of the light at different times is called Modal dispersion. It is also called modal distortion, multimode dispersion, intermodal distortion, pulse spreading and intermodal delay distortion.

Digital communication use light pulse to transmit signal down the length of the fiber. Modal dispersion causes pulses to spread out as they travel along the fiber, the more modes the fiber transmits, the more pulses spreadout. This significantly limits the bandwidth of step-index multimode fibers. The passage of light Step-index multimode fiber is shown in Figure 1.12.

Travels slower   Travels faster

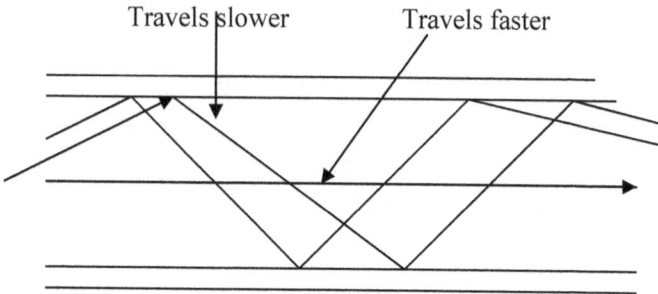

Figure 1.12   Passage of light Step-index multimode fiber

Graded-index multimode fibers solve the problem of modal dispersion. Graded-index fiber's refractive index decreases gradually away from its center, finally dropping to the same value as the cladding at the edge of the core. The change in refractive index causes refraction instead of total internal reflection which bends light rays back toward the fiber axis as they pass through the layers with lower refractive index. No total internal reflection happens because refraction bends light rays back into the fiber axis before they reach the cladding boundary.

Different light modes in a graded-index multimode fiber still follow different lengths along the fiber as in step-index multimode fiber. However, their speeds differ because the speed of guided light changes with fiber core's refractive index.

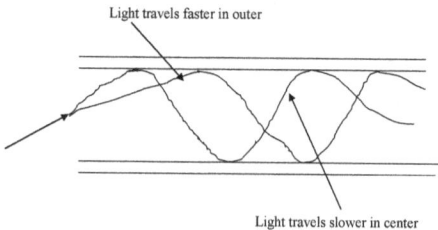

Light travels faster in outer

Light travels slower in center

Figure 1.13   Passage of light in Graded-index Multimode fiber

So, the farther the light goes from the center of the fiber, the faster its speed. So the speed difference compensate for the longer paths followed by the light rays that go farthest from the center of the fiber. This equalizing of transit times of different modes greatly reduces modal dispersion. The modal dispersion may be considered reduced but never completely eliminated. The passage of light in Graded-index Multimode fiber is illustrated in Figure 1.13.

For multimode propagation, dispersion is often expressed as a bandwidth length product (BLP) or bandwidth distance product (BDP). BLP indicates what signal frequencies can be propagated through a given distance of fiber cable and is expressed mathematically as the product of distance and bandwidth (sometimes called line width). Bandwidth length products are often expressed in MHz - km units. As the length of an optical cable increases, the bandwidth (and thus the bit rate) decreases in proportion. These bends become a great source of loss when the radius of curvature is less than several centimeters. Light propagating at the inner side of the bend travels a shorter distance than that on the outer side. To maintain the phase of the light wave, the mode phase velocity must increase. When the fiber bend is less than some critical radius, the mode phase velocity must increase to a speed greater than the speed of light. However, it is impossible to exceed the speed of light. This condition causes some of the light within the fiber to be converted to high-order modes. These high-order modes are then lost or radiated out of the fiber.

Intramodal dispersion

Intramodal dispersion is also called chromatic dispersion. Each component wavelength travels at slightly different speed in the fiber and hence it causes the light pulse to spread out as it travels along the fiber. This is known as chromatic dispersion which is the combination of material dispersion and waveguide dispersion. They change the transmission speed due to the atomic structure of the material and propagation characteristic of the fiber. Intramodal dispersion means the dispersion within a single mode. Although chromatic dispersion is associated with single mode fiber, it still occurs inmultimode fibers but the effect is reduced by the intermodal dispersion.

1.9.5 Signal to Noise Ratio

The performance of an optical communication system depends on the magnitude of the signal in comparison to the noise ratio. Signal to noise ratio (SNR or S/N) is calculated from the following formula

SNR = (peak to peak signal voltage)/(rms noise voltage)

In digital communication systems, a different figure of merit is also used. It is called the bit error rate (BER). The BER is the ratio of the number of wrong decisions made by the receiver to the total number of decisions made. That is,

BER = (number of bit errors) / (number of bits transmitted)

1.10 FIBER MAKING

In the earlier days, fibers were made from soda-lime silicate glasses. Corning glass works reported low loss glass fibers from mixtures of silicon dioxide and oxides of germanium, boron and other atoms using chemical vapour deposition technique, which is also known as outside chemical vapour deposition (OCVD) process. Subsequently, Bell Laboratories announced inside chemical vapour deposition process. Vapour axial deposition process was disclosed by NTT with Sumitomo, Fujikura and Furukawa companies. All these processes produce low loss high quality single mode and multi mode fibers. These processes are improved in the last two decades to increase the deposition rate of materials, the efficiency of use of starting materials, manufacturing cost and the best yield of the fabricated fibers.

Two methods are used to draw fiber directly. They are a) Double crucible method and b) Rod in tube method

1.10.1 Double Crucible Method

Double crucible method is a simple method and it is shown in Figure1.14. In this method, the molten core glass is placed in the inner crucible. The molten cladding glass is placed in the outer crucible. The two glasses come together at the base of the outer crucible and subsequently a fiber is drawn. Long fibers can be produced from this technique. Step-index fibers ad graded-index fibers can be drawn with this method.

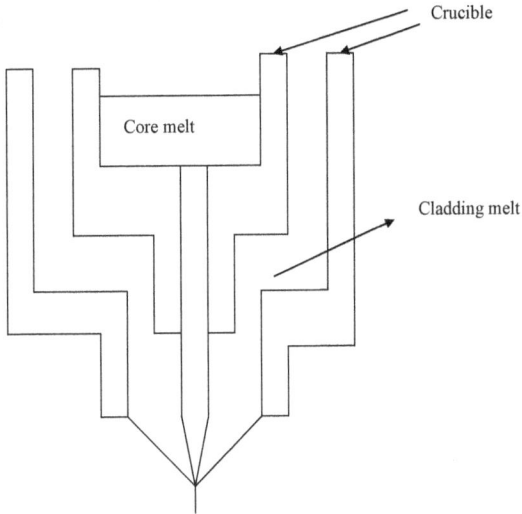

Figure 1.14   Double crucible apparatus

1.10.2 Rod in Tube Method

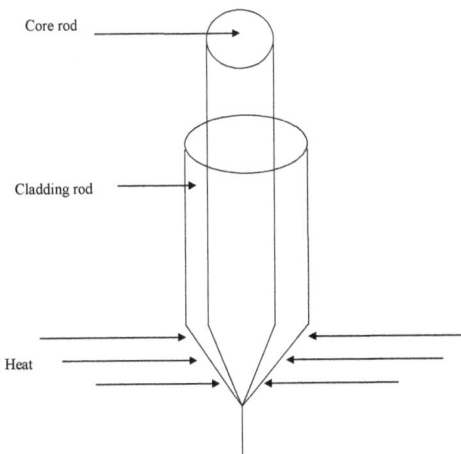

Figure 1.15   Rod in Tube apparatus

A rod of core glass is kept inside a tube of cladding glass. The end of this assembly is heated. When both the glasses are softened, glass fiber is drawn. The dimension of the rod and tube are usually 1 meter long. The core rod has typically a 30 mm diameter. It is important that the core glass and the cladding glass must have same or similar softening temperatures. The rod in tube apparatus is shown in Figure 1.15

This method is relatively easy provided if the rod and tube are available. However, one must be very careful not to introduce impurities between the core and the cladding.

1.10.3 Deposition Techniques

Most optical fibers are made from performs. The performs are made by depositing silica and various dopants from mixing certain chemicals. Once the perform is made, then fiber can be drawn from it.

In these techniques, oxygen and silicon tetrachloride react to make silica (SiO_2). Pure silica is doped with other chemical such as boron oxide, germanium dioxide and phosphorous pentoxide. They are used to change the refractive index of the glass.

Many techniques are used to make performs. They are

1.   Modified chemical vapour deposition technique (MCVD)

2.   Plasma enhanced modified chemical vapour deposition technique (PMCVD)

3.   Outside vapour deposition (OVD)

4.   Inside vapour deposition (IVD)

5.   Axial vapour deposition (AVD)

1.10.4 Modified Chemical Vapour Deposition Technique (MCVD)

The chemicals are mixed inside a glass tube which rotates on a lathe. They react and extremely fine particles of germano or phosphoro silicate glass are deposited on the inside of the tube. A moving burner moves along the axis of tube. This causes chemical reaction and fuses the deposited material.

The perform is deposited layer by layer by starting first with the cladding layers and followed by the core layers. Varying the composition of chemicals changes the refractive index of the glass. When the deposition is completed, the tube is collapsed at 2000°C into a perform of the purest silica with core of different composition. The Block diagram for Modified Chemical Vapour Deposition (MCVD) system is shown in Figure 1. 16.

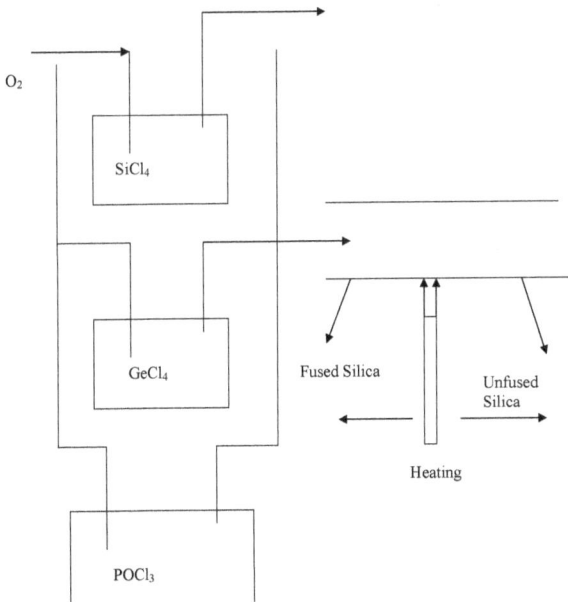

Figure 1.16 Block diagram of Modified Chemical Vapour Deposition (MCVD) system

1.10.5 Plasma Enhanced Modified Chemical Vapour Deposition Technique (PMCVD)

Plasma enhanced modified chemical vapour deposition technique is similar to MCVD in principle. However, the difference is due to the use of plasma instead of torch. The fourth state plasma is nothing but electrically heated ionized gases. It provides sufficient heat to increase the chemical reaction rates inside the tube and the deposition rate. This technique can be used to manufacture very long fibers of the order 50 Km. This technique is used for producing step-index and graded-index fibers. The block diagram of Plasma Enhanced Modified Chemical Vapour Deposition (PMCVD) system is shown in Figure 1. 17.

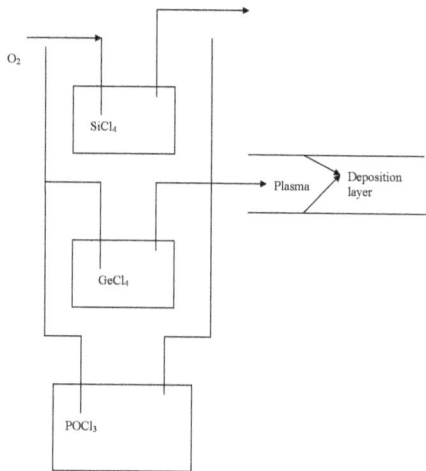

Figure 1.17 Block diagram of Plasma Enhanced Modified Chemical Vapour Deposition (PMCVD) system

1.10.6 Outside Vapour Deposition Technique (OVD)

In this technique the chemical vapours are oxidized in a flame in a process known as hydrolysis and the deposition is done on the outside of a silica rod. Very pure SiO_2 tiny particles are obtained by passing oxygen to pure silicon tetrachloride liquid. These particles are directed to a rotating mandrel to deposit in the form of soot. The torch moves back and forth (laterally) over the mandrel, a 'soot perform' is build up. When sufficiently large soot perform has been deposited, the perform is placed in a furnance for sintering. Finally, the target mandrel is removed and the resulting tube is thermally collapsed into a solid rod. Fiber can be drawn from this solid rod. Figure 1.18 shows the block diagram of Outside Vapour Deposition (OVD) system.

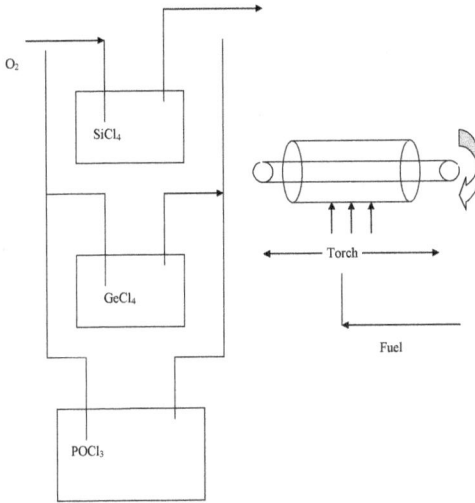

Figure 1.18 Block diagram of Outside Vapour Deposition (OVD) system

1.10.7 Axial Vapour Deposition Technique (AVD)

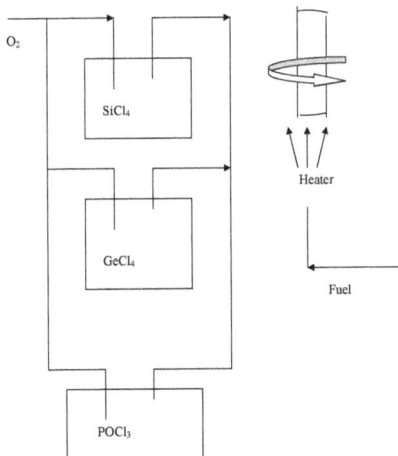

Figure 1.19 Block diagram of axial vapour deposition (AVD) system

In this technique, the deposition occurs on the end of a rotating silica boule as chemical vapours react to form silica. The resulting soot perform is sintered into a solid rod. Core performs and very long fibers can be made with this technique. Step-index fibers and graded-index fibers can be manufactured using this technique. The block diagram of Axial Vapour Deposition (AVD) system is shown in Figure 1.19.

1.10.8 Fibers from Preform

All the deposition techniques listed above produce preforms. Preforms are typically 1 m long with 2 cm diameter. Manufacturers of optical fibers vary this dimension as per the needs. Preform is the basis to produce thin optical fibers. The process to produce fibers is known as drawing. The typical fiber drawing apparatus is shown in Figure 1. 20.

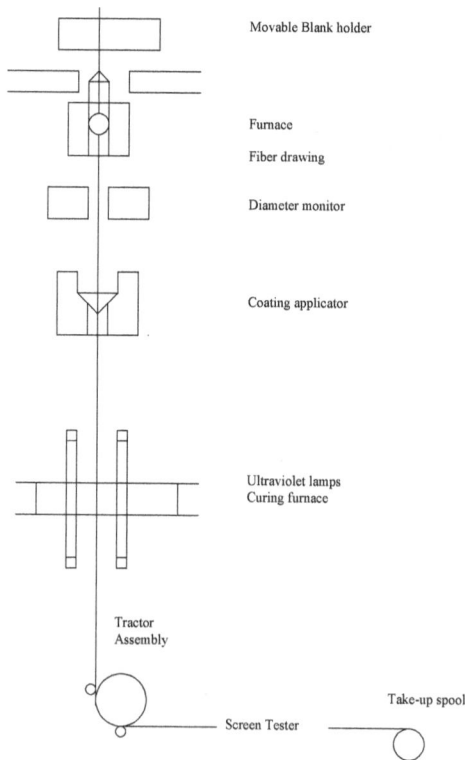

Figure 1. 20   Fiber drawing and spooling

In this final step of the fabrication process, fiber undergoes several treatments. After the fiber is drawn from the preform, it undergoes quality checking. It is also coated for protection. Its diameter is accurately monitored and finally it is stored on a spool. The tip of the preform is heated in an oven at a controlled rate to about 2000°C. As the glass softens, a thin strand of softened glass falls by gravity and cools down. As the fiber is drawn, its diameter is constantly monitored. It is necessary to control the fiber outside diameter to facilitate splicing and cabling. A Plastic coating is than applied to the fiber before it touches any components. The coating protects the fiber from abrasion, dust and moisture. As a last step, the fiber is then wrapped around a spool. At this stage, affixed amount of strain is applied (proof stress) on the fiber to confirm that the fiber is string enough to use in actual cables.

REVIEW QUESTIONS

1. Distinguish between Rayleigh, Mie and Raman scattering.
2. List the advantages of optical fibers.
3. What is the necessity for fiber optics?
4. Explain the principle of light propagation through optical fiber.
5. Describe the basic structure of an optical fiber.
6. Write a note on
   a. numerical aperture,
   b. acceptance angle.
7. Describe a single mode fiber. What are the advantages and disadvantages of a single mode fiber?
8. Explain the modes in a multimode and single mode fiber.
9. Describe a multimode fiber and light transmission in it.
10. Distinguish between step index and graded index multimode fiber.

11. What is meant by optical fiber index profile? Explain.

12. What is attenuation loss? List the different types of attenuation loss in optical fiber and explain each one of them in detail.

13. Distinguish between inter-modal and intra-modal dispersion.

14. What do you understand by signal to noise ratio?

15. With necessary diagram, describe fiber making technique using

    i.   double crucible technique and

    ii. rod on tube technique.

16. What is deposition technique? Explain with neat diagrams, all the deposition techniques in fiber making process.

17. How the optical fibers are obtained from perform?

Chapter 2

OPTICAL SOURCES AND DETECTORS

To make optical fiber communication successful, sources and detectors should be efficient. These sources and detectors have been developed by taking advantages in semiconductor devices. Light emitting diodes are used in short distance communication, while laser diodes are preferred for long distance communication. PIN diodes and avalanche diodes are used as detectors. Due to the rapid development in semiconductor devices, we can expect breakthrough in diode lasers and detectors in the coming years.

2.1 TYPES OF MATERIALS

There are three important classes of materials via, conductors, insulators and semiconductors.

1. Conductors such as metals have the best conductivities because they have the greatest number of charge carriers per atom. Typical metallic conductors such as silver have about one free electron per atom which gives them a much higher conductivity than the semiconductors.

2. Insulators have very nearly all their electrons tightly bound to their atoms in their lattices with very few free to move for conduction. A good insulator might have one charge carrier for each 10^{20} or more atoms.

3. Semiconductors have conductivities intermediate between insulators and conductors. Pure gallium arsenide at room temperature has less charge carrier while in doped germanium has more charge carrier for conduction. These carriers are free to move through their lattices under the influence of an electric field and thus carry an electric current. In semiconductors, conductivity can be varied over orders of magnitude by changes in temperature, optical excitation and doping. The elemental semiconductor Si, Ge and compound semiconductors such as GaAs play a vital role in several applications.

Semiconductors are used in the integrated circuits SSI, MSI, LSI and VLSI and group II–VI semiconductors such as ZnS are used as fluorescent materials in TV screens. Compound semiconductors such as InSb, CdSe, PbTe and HgCdTe are used as light detectors while infrared and nuclear radiation detectors use Si and Ge. The microwave device uses gunn diode. The compound semiconductors GaAs, GaP, InP and AlGaAs are used in LEDs and laser diodes.

2.2 LIGHT EMITTING DIODES

In light emitting diodes, light is produced by a solid state process called electroluminescence. LEDs life span is around 100,000 hours compared to about 1000 hours for an incandescent bulb. Recently, blue LEDs have become a reality due to GaN while white light LEDs can be produced by combining the red, green and blue chips in a single device. The structure of LED the circuit connection is shown in Figure 2.1.

LEDs wavelengths are the peak output, which is expressed in nanometers. LEDs are not perfectly monochromatic but they produce wavelengths over a small region of the spectrum. Spectral line half-width is the width of the curve at 50% intensity and is a measure of the purity (monochromaticity) of the color. LEDs emit slightly different colors at different temperatures. They also emit different colors at different currents. White LEDs which depend on phosphors to change the coloured light of the dye to white light. Infrared LEDs are sometimes called IREDs (Infra Red Emitting Diodes). 400 nm is a pretty common wavelength for UV LEDs. The powers of the LEDs are rated in milliwatts.

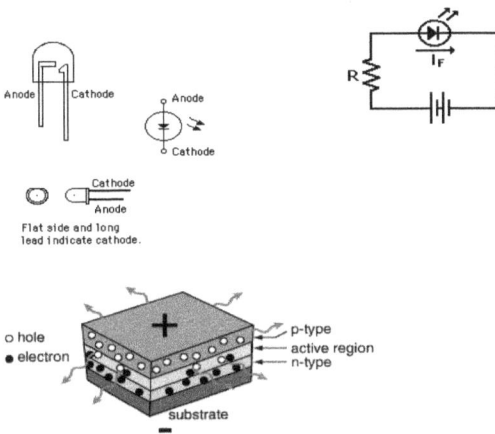

Figure 2.1   Structure of LED the circuit connection

A light emitting diode has essentially the same structure as a laser diode but has essentially zero reflectivity facets and therefore never reaches a lasing threshold. An LED is constructed by depositing three semiconductor layers on a substrate. An active region lying in between *p*-type and *n*-type semiconductor layers emits light when an electron and hole recombine. This is a solid state process known as electroluminescence. Since candela and lumen are units of measure used for visible-light LEDs, they can't be used for UV and IR LEDs. IR and UV LEDs are measured in watts for radiant flux and watts/steradian for radiant intensity. A fairly typical "bright" IR LED put out about 27 mW/sr, though they go up to 250 mW/sr or so. Signaling LEDs, like for TV remotes, are considerably less powerful. As a rule of thumb, different color LEDs requires different forward voltages to operate. For example, a red LEDs take the least voltage while the blue LEDs require increased voltage. Typically, a red LED requires about 2 volts while blue LEDs require around 4 volts. Typical LEDs, however, require 20 to 30 mA of current, regardless of their voltage requirements. A typical I – V characteristic curve for a LED is shown in Figure 2.2.

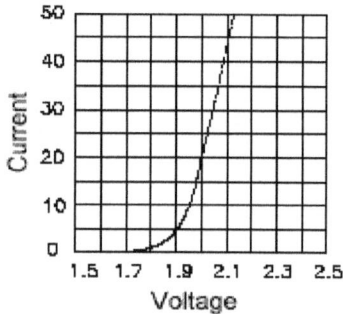

Figure 2.2   I—V curve for a LED

From Figure 2.2, it is seen that LED draws *no* current under 1.7 volts which is known as OFF position. Between 1.7 volts and about 1.95 volts, the "dynamic resistance" (the ratio of voltage to current) of LED decreases to 4 ohms. Above 1.95 volts, the LED is fully "on" and dynamic resistance remains constant. Ohm's Law doesn't work for LEDs due to non-linear relationship between voltage and current. LEDs have a much more vertical slope than do normal diodes. That is a tiny increase in voltage can produce a large increase current. When LED requires 2 volts for proper conduction, just as little as 2.04 volts could destroy it. To keep the current down to a reasonable level, a series resistor must be included in the circuit.

The formula for calculating the value of the series resistor is

$$R_{series} = (V - V_f) / I_f$$

where, R_{series} is the resistor value in ohms, V is the supply voltage, V_f is the voltage drop across the LED, and I_f is the current in the LED.

When LED is forward biased to the threshold of conduction, its current increases rapidly and must be controlled to prevent destruction of the device. The light output is quite linearly proportional to the current within its active region, so the light output can be precisely modulated to send an undistorted signal through a fiber optic cable. The Figure 2.3 depicts the forward current and light output.

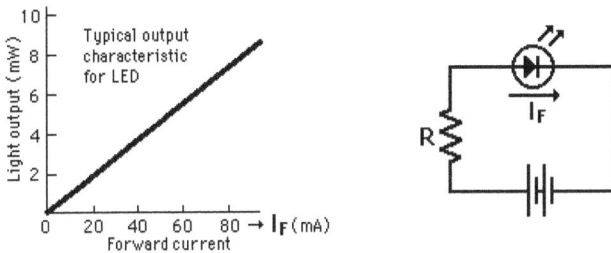

Figure 2.3 The forward current and light output.

Compared with the laser, the LED has a lower power output, slower switching speed and higher spectral width which lead to more dispersion. These deficiencies make the LED inferior for use with high speed data links and telecommunications. However, it is widely used for short and medium range communications using both glass and plastic fibers which are cheap, reliable, simple and is less temperature dependent. Further, it is unaffected by the incoming light energy from Fresnel reflection.

2.3 LASERS

2.3.1 Basics of Lasers

The lowest energy in atom is the ground state and other states are excited states. Under ordinary conditions, almost all atoms and molecules are in their ground states. Three types of processes are possible for a two-level atomic system. In the first, an incoming photon excites the atomic system from a lower energy state into a higher energy state. This is called absorption. The absorption depends on the population difference between lower level N_1 and higher N_2 and the refractive index of the medium.

When the atom or molecule has been lifted to excited state, there is a probability that it will emit radiation (photon) again while returning to a lower energy state. This lower energy state may be either the ground state or still one of the excited states but having lower energy level. The emission of a photon is called spontaneous emission or fluorescence. This emission process is random and incoherent.

Excited atoms can loose their energy not only by spontaneous emission but also by induced or stimulated emission. The stimulated emission has phase relations with each other and the output is coherent. Rate of stimulated emission, R_{21} (stim), from level 2 to 1 is given by

$$R_{21}\left(\text{stim}\right) = B_{21} N_2 \rho \qquad (1)$$

where, B_{21} is the Einstein's coefficient for stimulated emission and has the dimensions m^3/s^2J. N_2 is the population in the excited state and ρ is the radiation energy density. For an ideal material with only two non-degenerate energy levels,

The rate of spontaneous emission, R_{21} (spon), from level 2 to 1 is given as:

$$R_{21} \text{ (spon)} = A_{21} N_2 \qquad (2)$$

Absorption = spontaneous emission + stimulated emission

$$\text{i.e. } B_{12} N_1 \rho = A_{21} N_2 + B_{21} N_2 \, \rho \qquad (3)$$

At any given instance, under normal circumstances, both stimulated and spontaneous emissions may occur but the probability of stimulated emission is pretty low. One can find out this ratio of spontaneous to stimulated emission using one of the following equations.

$$\left(\frac{R_{spon}}{R_{stim}}\right) = \left(\frac{A_{21}}{\rho B_{21}}\right) = \exp\left(\frac{h\nu}{kT}\right) - 1 \tag{4}$$

$$\frac{R_{21}(spon)}{R_{21}(stim)} = \frac{8\pi h\nu^3}{c^3\rho} \tag{5}$$

The ratio of the probability of spontaneous to stimulated light emission depends directly on the frequency of emission or inversely to the wavelength. In the optical region, spontaneous emission occurs commonly in contrast to stimulated emission.

In a two level system, the number of atoms in the excited state can never exceed the number in the ground state and hence two level lasers are not possible. The laser action is possible only when $N_2 > N_1$. This non-equilibrium condition is known as population inversion.

Although population inversion is the primary condition for laser action, it is not sufficient for producing a laser. Added to spontaneous emission, emitted photon undergoes certain losses. This has to be compensated through the geometry of the system so that one can achieve overall gain. Hence, we require an optical cavity or a resonator.

By population inversion, a large number of atoms are created in excited states. If one of the atoms emitted spontaneously, then the emitted photon would stimulate other atoms to emit. These emitted photons in turn stimulate further emission. The result is amplification with intense burst of coherent radiation.

For the laser to operate, we need

1. An active medium with a suitable set of energy levels to support laser action.
2. A source of pumping energy in order to establish a population inversion and
3. An optical cavity or resonator to introduce optical feedback. This serves to maintain the gain of the system overcoming all losses.

Optical resonator plays a very important role in the generation of the laser output, in providing high directionality to the laser beam as well as producing gain in the active medium. Hence, the losses due to the photons away from the laser medium, diffraction losses due to the mirrors, radiation losses inside the active medium due to absorption and scattering etc. are compensated. In order to sustain laser action, one has to confine the laser medium and the pumping mechanism in a special way that should promote stimulated emission rather than spontaneous emission. In practice, photons need to be confined in the system to allow the number of photons created by stimulated emission to exceed all other mechanisms. This is achieved by bounding the laser medium between two mirrors. On one end of the active medium is the high reflectance mirror (100% reflecting) or the rear mirror and on the other end is the partially reflecting or transmissive mirror or the output coupler. The laser emanates from the output coupler, as it is partially transmissive. Stimulated photons can bounce back and forward along the cavity, creating more stimulated emission as they go. In this process, the photons not traveling along the optical axis as well as the photons of incorrect frequency are lost.

2.3.2 Types of Lasers

Today, we are familiar with lasers such as Helium–Neon and diode lasers as bar code reader in shops, supermarkets, in home CD players, in laser printers and in hundreds of alignment tasks. The number of laser system in use today differs in active medium and in pumping methods. A brief list of lasers used in various applications are presented in Table 2.1.

Table 2.1 Lasers and their applications

Sl.No	Lasers	Wavelength(nm)
	SOLID STATE LASERS	
1.	Holmium:YAG (Ho:YAG)	2100 (mid–IR)
2.	Neodymium:YAG (Nd:YAG)	1318 (mid–IR)
3.	Neodymium:YAG (Nd:YAG)	1064 (near IR)
4.	Ruby	694
5.	Diode lasers	Variable with system
6.	Erbium: YAG (Er:YAG)	2940 (mid IR)
7.	KTP (Potassium Titanyl phosphate)	532
	GAS LASERS	
8.	Carbon–dioxide (CO_2)	10600 (far–IR)
9.	Hydrogen fluoride (HF)	2040 (mid– IR)
10.	Krypton	647, 568 and 531
11.	Helium–Neon (He–Ne)	632
12.	Gold Vapour	632
13.	Copper Vapour	577 and 510
14.	Argon	488 and 515
15.	Excimer	UV region
	ArF	193
	KrCl	222
	KrF	248
	XeCl	308
	XeF	351
	LIQUID LASERS	
16.	Tunable dye lasers	632, 577 and 504

2.3.3 SOLID STATE LASERS

Solid state lasers are made by a crystal or a glass with small amount of doping such that the upper laser level is optically pumped by a suitable light source. There is another class of laser using a solid as active medium which are becoming more and more important. It is a semiconductor crystal, usually with one or more junctions and pumping is achieved by injecting a current in the crystal. The solid state laser (SSL), also known as OPTICAL MASER, was the first laser operated in 1960 by Maiman. He obtained laser action in a ruby crystal ($Cr^{3+}:Al_2O_3$), excited by a pulsed flashlamp. Today SSLs are still one of the most important classes of lasers, both in industrial applications, where the only competitor in average power is the CO_2 laser, as well as in medical applications where the possibility to transport the laser beam by a thin optical fiber opens the possibility of easy surgery.

In recent years, the pumping mechanism is done not by flashlamps but by semiconductor lasers. Hence, we have a new class of lasers pumped by diodes. They are known as diode pumped solid state lasers.

Diode pumped solid state lasers (DPSSL) In the history of lasers, 1980 is a remarkable year. In this year, an efficient, powerful room temperature AlGaAs semiconductor laser revolutionised the field of solid state lasers. Replacement of conventional flashlamps paved a path for diode pumped solid state lasers. Many Nd^{3+} ions doped laser crystals such as $Nd:LiYF_4$ and $Nd:YVO_4$ have emerged. As you know, Erbium doped fiber amplifier (EDFA) needed 980 nm laser to make a revolution in amplifying optical signals. Luckily, InGaAs laser diode with 980 nm was developed. This laser was used as diode pumped laser for Yb:YAG laser which was discarded as inefficient laser during 1960s. Hence, Ytterbium (Yb^{3+}) doped solid state laser (Yb:YAG) pumped with InGaAs laser diodes has been intensively and successfully developed. A search is in progress for novel Yb doped crystals possessing properties superior to known Yb laser with new capabilities. Several such materials have been identified and characterised recently. This sub- heading concerns with such new laser materials for diode pumped solid state lasers and its potential use in cosmetic, medical and veterinary photonics.

Photonics is the technology of generating and harnessing light and the other forms of radiant energy whose quantum unit is photon. This Century is portrayed as the photonics era. Basic physics provides insights into how we can use PHOTONICS TECHNOLOGY to understand and change the world around us. The range of applications of photonics extends from energy generation to communication, information processing and a variety of tasks beneficial to the society. In recent years, great interest has been shown in photonics for medical applications because of their useful emission wavelength. Initially, use of lasers in medicine was initiated mostly by natural scientific curiosity and hoped that the laser could become a super tool. Today the situation has completely changed. More delicate and precise lasers are available at present for skilled surgeons to perform even complicated and delicate operations. The lasers have now shifted to the hands of plastic surgeons and beauticians. Such types of lasers are widely known as cosmetic lasers.

DEVELOPMENTS IN DIODE PUMPED SOLID STATE LASERS

The recent advances in Diode pumped solid state lasers (DPSSL) made a revolution in laser Physics. DPSS lasers in many ways approach the ideal output power, beam quality and repeatability to those of a gas lasers but their efficiency and size are more like those of semiconductor diode lasers. Although the initial cost of a DPSSL and an ion laser are comparable, annual operating cost of the DPSSL system is significantly less because of its operating lifetime of nearly 10,000 hours. With modern InGaAs diode pumps, Yb:YAG lasers have been intensively developed and today we have successful Yb:DPSSLs. Yb based lasers are said to operate in a "quasi three level" laser scheme. Beyond Yb:YAG, attempt has been made to achieve Yb doped crystals with highly favourable spectroscopic properties. The nomenclature for new Yb doped laser materials are

NOMENCLATURE

DPSSL	Diode Pumped Solid State Laser
YAG	$Y_3Al_5O_{12}$
FAP	$Ca_5(PO_4)_3F$
S–FAP	$Sr_5(PO_4)_3F$
KGW	$KGd(WO_4)_2$
KYW	$KY(WO_4)_2$
BCBF	$BaCaBO_3F$
CAS	$Ca_2Al_2SiO_7$
LNB	$LiNbO_3$
LTA	$LiTaO_3$
YCOB	$YCa_4O(BO_3)_3$

Subsequently, materials science research yielded several Yb doped crystals including hosts.

$LiNbO_3$ – (LNB)
$Ca_2Al_2SiO_7$ – (CAS)
$BaCaBO_3F$ – (BCBF)
$SrLaGaO_4$
$SrLaGa_3O_7$
$KY(WO_4)$ – (KYW)
$KGa(WO_4)_2$ – (KGW)
$CuGaS_2$
$AgGaSe_2$
$AgGaS_2$
$LiTaO_3$ – (LTA)
$YCa_4O(BO_3)_3$ – (YCOB)

Yb doped crystal DPSSLs continue to advance in performance using well developed Yb:YAG crystal and developing Yb doped fluorapatite crystals. Yb doped double tungstates (KYW &

KGW) provide a new opportunity for design of practical Yb based DPSSLs. A continued research is necessary for Yb based laser crystals amenable to channel waveguide formation and possessing more favourable laser spectroscopic values.

The novel self frequency doubling Yb:YCOB and Yb:GdCOB crystals seem destined for practical use in visible DPSSL applications. A 10 W Q-switched Ultra violet DPSSL sets a new record in operating at 355 nm.

Diode pumped solid state laser sources started replacing gas and ion laser sources for commercial and laboratory applications, and the trend is accelerating because of progress in two areas.

First, more powerful, reliable laser diodes are available at lower cost, making diode pumping not only attractive but also practical.

Secondly, advances in manufacturing have improved the quality while reducing the volume pricing for laser crystals such as Nd:YAG and Nd:YVO$_4$, and nonlinear optical crystals such as β–BaB$_2$O$_4$ (BBO), LiB$_3$O$_5$ (LBO) and KTiOPO$_4$ (KTP). To meet requirements of expanding laser applications, important progress has occurred in searching out new crystals that can cover broader spectral applications.

Non-linear optic crystals The non-linear crystals are important because the fundamental output wavelength of Nd:YAG and Nd:YVO$_4$ is 1064 nm in the near infrared, a wavelength that is extremely useful for many applications. However, if these lasers are to replace gas and ion lasers that emit in the visible and ultra violet portions of the spectrum, they need some type of device that can be shortened the wavelengths by doubling or tripling their frequencies.

As the most powerful laser frequency conversion devices, nonlinear optical crystals are playing increasingly important role in advanced laser technologies. The development of new crystals has combined with other laser technologies (Diode pumping, Ultra fast technologies and Optical parametric processes) to greatly accelerate the creation of all solid state lasers. Because of these changes, recently introduced laser sources have much shorter wavelengths (down to 187 nm), faster laser pulses (down to less than 10 fs), much wider tuning ranges (from 187 nm – 10 μm) and higher powers and yet are considerably smaller than the predecessors.

Several low power (10–400 mW) diode pumped solid state green lasers with performances similar to the air-cooled argon ion lasers are available at present and they are typically used in graphics and diagnostic medical applications. The high power pulsed lasers are used in micro machining, marking and trimming applications.

For UV generations (third, fourth and fifth harmonics of Nd:YAG and Nd:YVO$_4$ lasers), output power has reached to 1–2.5 W. BBO and LBO have proved to the only candidates that could generate high power over a considerable period. BBO will continue to be the most important crystal for deep-UV generation.

Meanwhile, there is an increasing demand for all solid state, high power, deep UV laser sources for fiber Bragg grating fabrication, laser drilling of polymers and other industrial applications. Super–BBO (S–BBO) crystal is for rescue, which offers high transmission at 266 nm, low scattering and good optical homogeneity.

The most promising non-linear crystals are:

1. For the far ultraviolet, Sr2Be2B2O7 (SBBO), KBe2BO3F2 (KBBF).

2. For the infrared, KTiOAsO4 (KTA), ZnGeP2, AgGaSe2/ AgGaS2.

3. Periodically pooled nonlinear crystals such as LiNbO3 (PPLN) and KTP (PPKTP) have demonstrated very low operating thresholds and high efficiencies for optical parametric and Harmonic operations. The compact size of the system due to non-linear crystals has made many applications possible.

In general, materials for laser operation must possess sharp fluorescent lines, strong absorption bands, and reasonably high quantum efficiency for the fluorescent transition of interest. These characteristics are often shown by solids such as crystals and glasses. A very small quantity doping elements viz., the rare earths namely lanthanides and actinides, in them causes optical transitions between states of inner incomplete electron shells. The sharp fluorescence lines in the spectra of crystals doped with these elements result from the fact that the electron involved in transition in the optical regime are shielded by the outer shells from the surrounding crystals lattice. The corresponding transitions are similar to those of the free ions. In addition to a sharp fluorescence emission line, a laser material suitable for optical pumping should possess broad-band pump transitions i.e., incandescent lamps, CW arc lamps, or flashlamps are used as pump sources for pumping solid state lasers. The gain in solid state lasers is achieved through

a. The host material with macroscopic (mechanical, thermal and optical) and microscopic (lattice) properties,

b. The active ions with their distinctive charge states and free ion electronic configurations and

c. The optical pump source with particular configuration, spectral irradiance and duration.

These elements are interactive and its proper selection yield high performance laser.

Solid state host materials may be broadly grouped into crystalline solids and glasses. The host must have good optical, mechanical and thermal properties to withstand the severe operating conditions of lasers. The desirable properties of the host material include hardness, chemical inertness, absence of internal strain and refractive index variation, resistance to radiation induced color centers and ease of fabrication. Several interactions between the host crystal and the active ion restrict the number of useful material combinations. These include size, disparity, valence and spectroscopic properties. Ideally, the size and valence of the additive ion should match with that of the host ion which it replaces.

GLASSES Glasses are important class of host materials for some of the rare earths, particularly Nd and very recently Yb. The outstanding practical advantage compared to crystalline materials is the tremendous size capability for high-energy applications. Rods up to one meter in length and over 10 cm in diameter and disks up to 90 cm in diameter and several cm in thickness are currently available for laser hosts. The optical quality is excellent and beam divergence approaching the diffraction limit can be achieved in glasses. Glass can be easily fabricated and yields a good optical finish. Laser ions placed in glass generally show a large fluorescent line-width than in crystals as a result of the lack of a unique and well-defined crystalline surrounding for the individual active atom. Therefore, the laser threshold for glass lasers has been found to run higher than their crystalline counterparts. Further, glass has a much lower thermal conductivity than most crystalline hosts. The latter factor leads to thermally induced birefringence and optical distortion in glass rods when they are operated at high average powers. For these reasons, glass media are used mainly in high energy laser systems with low repetition rate. In addition to the Nd:glass, another important glass laser medium is Er:glass. Glass doped with erbium is of special importance because its radiation 1.55 μm does not penetrate the lens of the human eye, and therefore cannot destroy the retina. Due to the three-level behaviour of erbium and its characteristic small light, the multiple doping with neodynium and ytterbium is necessary to obtain satisfactory system efficiency. Ho:glass is another laser medium which attracted the medical profession for medical lasers.

CRYSTALS A large number of crystalline host materials have been investigated since the discovery of the ruby laser. Crystalline laser hosts are advantageous over glasses due to their higher thermal conductivity, narrower fluorescence line-widths and in many cases, greater hardness. For all these reasons, they are used for high average power, CW or repetitively pulsed lasers. However, the optical quality and doping homogeneity of crystalline hosts are often poorer and the absorption lines are generally narrower. Therefore, let us review briefly two typical crystals mainly used as laser hosts.

SAPPHIRE The first laser material Sapphire was used as a host in ruby laser. The Al_2O_3 (Sapphire) host is a hard and high thermal conductivity material. The transition metals can readily be substituted for the Al. The Al site is too small for rare earth and it is not possible to incorporate appreciable concentrations of these impurities into sapphire. Besides, ruby and Ti-doped sapphire has gained significance as a tunable laser material.

GARNETS Some of the most useful laser hosts are the synthetic garnets such as aluminium garnet $Y_3Al_5O_{12}$ (YAG), gadolinium gallium garnet, $Gd_3Ga_5O_{12}$ (GGG) and gadolinium scandium gallium garnet $Gd_3Sc_2Ga_3O_{12}$ (GSGG). These garnets have many desirable properties to use it as laser host material. They are stable, hard, optically isotopic and have good thermal conductivities which permit laser operation at high average power levels. In particular, yttrium aluminium garnet doped with neodymium (Nd:YAG) has achieved a high status among solid-state laser materials. In recent days, ytterbium doped yttrium aluminium garnet is becoming popular due to the ytterbium's quasi three level. YAG is a very hard, isotropic crystal which can be grown and fabricated in a manner that yields rods of high optical quality. At the present time, it is the best commercially available crystalline laser host for Nd^{3+}, (although several new host materials are available at present) offering low threshold and high gain. The most important ions used in crystal doping for SSLs can be divided into two main classes viz, transition metal ions and rare earth ions. The most important members of the transition metal family are Cr^{3+} and Ti^{3+}. The former was the first to show laser action, using Sapphire as host crystal (ruby:Cr^{3+}:Al_2O_3).

Titanium used in sapphire shows a very broad tunable emission range which is useful for spectroscopic applications. The rare earth ions are the natural candidates to serve as active ions in solid-state laser materials because they exhibit a wealth of sharp fluorescent transitions representing almost every region of the visible and near-infrared portions of the electromagnetic spectrum.

These lines are very sharp, even in the presence of the strong local crystal fields. The other rare earth ions have also shown laser action in many different crystals. Out of several rare earth ions, Nd^{3+} was used as a common dopant for many crystals as well as in glasses. Nd^{3+}, Er^{3+}, Ho^{3+} and Yb^{3+} are becoming popular day-by-day due to their eye safe emission wavelength.

NEODYMIUM Nd^{3+} was the first trivalent rare earth ions to be used in a laser and it remains as the most important element in this group. Stimulated emission has been obtained with this ion by incorporating it with several host materials. Further, a high power level has been obtained from Nd lasers than from any other four-level material. The principal host materials are YAG and glass. In these hosts, stimulated emission is obtained at a number of frequencies within three different groups of transitions centered at 0.9, 1.06 and 1.35 μm.

ERBIUM Numerous studies of the absorption and fluorescence properties of erbium in various host materials have been conducted to determine its potential as an active laser ion. Laser oscillation was observed most frequently in the wavelength region 1.53 to 1.66 μm. Stimulated emission around 1.6 μm is of interest to ophthalmologists because the eye is subjected to less retinal damage by laser radiation at 1.6 μm. The reduced transmissivity of the ocular media is the reason for selecting this wavelength.

LASER CRYSTALS Among the lot of crystals used as solid state lasers active media, a few have gained great prominence. The first is ruby (Cr^{3+} in sapphire), both for historical reasons (it was the first laser to be operated) as well as for some unique properties. Next comes the Nd: YAG and all the other Nd based materials. Recently, a new class of materials viz., Alexandrite and Ti:Sapphire has become popular and they can be used as the tunable laser crystals.

RUBY The ruby laser, although a three-level system, still remains in use today for many applications. From an application point of view, ruby is attractive because its output lies in the visible range, in contrast to most rare earth four-level lasers, whose outputs are in the near-infrared region. Photo detectors and photographic emulsions are much more sensitive at the ruby wavelength than in the infrared. Spectroscopically, ruby possesses an unusually favourable combination of a relatively narrow line-width, long fluorescent lifetime, a high quantum efficiency, broad and

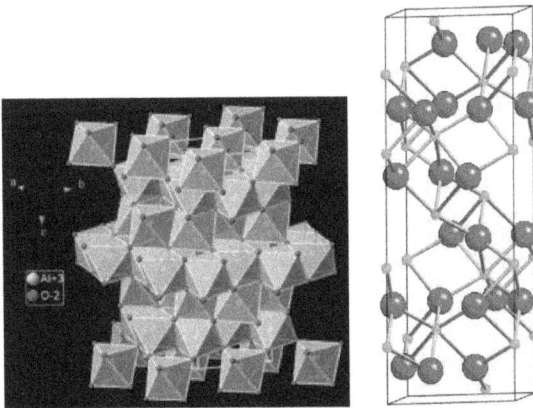

Figure 2.4 Structure of Al_2O_3 (sapphire/corundum) [a = b = 4.7617 [Å]; c = 12.9947 [Å]; $\alpha = \beta = 90^0$, $\gamma = 120^0$]

well-located pump absorption bands which make unusually efficient use of the pump radiation emitted by available flashlamps. Ruby chemically consists of sapphire (Al_2O_3) in which a small percentage of the Al^{3+} has been replaced by Cr^{3+}. This is done by adding small amounts of Cr_2O_3 to the melt of highly purified Al_2O_3. The pure single host crystal is uniaxial and possesses a rhombohedral or hexagonal unit cell as shown in Figure 2.4.

The crystal has an axis of symmetry, namely c axis, which forms the major diagonal of the unit cell. Since the crystal is uniaxial, it has two indices of refraction, the ordinary ray having the E vector perpendicular to the c (optic) axis, and the extraordinary ray having the E vector parallel to the c axis. As a laser host crystal, sapphire has many desirable physical and chemical properties. The crystal is a refractory material, hard and durable. It has a good thermal conductivity, chemical stability, and is capable of being grown by the Czochralski method to very high quality. In this procedure, the solid crystal is slowly pulled from a liquid melt by initiation of growth on high-quality seed material. Iridium crucibles and radio frequency heating are used to contain the melt and control the melt temperature, respectively. The crystal boles can be grown in the 0°, 60° or 90° configuration, where the term refers to the angle between the growth axis and the crystallographic c axis. For laser grade, 60° ruby type is commonly used. Ruby can be grown in relatively larger boules with high optical quality. The boules are cut into smaller cylindrical sections. Core-free cylindrical rods can be obtained up to 30 cm in length and 2.5 cm in diameter. Typical commercially available rods with flat ends are fabricated to the following specifications.

a.   Ends flat to l/10,

b.   Ends parallel to growth axis (0°, 60°, 90°) to +5.

c.   Common rod geometries may be plane parallel, Wedged surfaces, Brewster angle, or prismatic.

d.   The best way to inspect a laser rod is by means of interferometer which shows optical inhomogeneities, strain and distortion, etc., as fringes. Scattering centers of regions of high defects in a laser rod are revealed by illuminating the crystal from the side with a He–Ne gas laser. The energy level diagram of ruby is shown in Figure 2.5.

Ruby is a synthetic crystal of aluminum oxide (Al_2O_3) and is more familiar in daily life as a precious stone for jewel. The chemical structure of Ruby is of Al_2O_3 (which is called Sapphire), with impurity of about 0.05% (by weight) of Chromium Ions (Cr^{+3}). The active ion is Cr^{+3}, which replace Al atom in the crystal. This ion causes the red colour of the crystal. The impurity ion of Cr^{+3} is responsible for the energy levels which participate in the process of lasing. This system is a three level laser with lasing transitions between E_2 and E_1. The excitation of the Chromium ions is done by light pulses from flashlamps (usually Xenon).

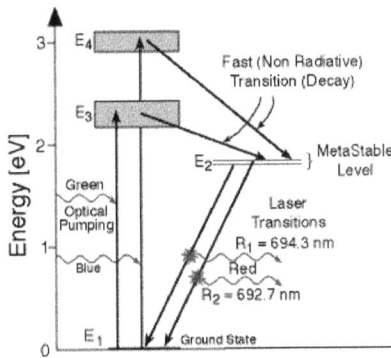

Figure 2.5 Energy Level Diagram of a Ruby Laser

The Chromium ions absorb light at wavelengths around 545 nm. As a result, the ions are transferred to the excited energy level E_3. From this level, the ions are going down to the metastable energy level E_2 in a non-radiative transition. The energy released in this non-radiative transition is transferred to the crystal vibrations and changed into heat that must be removed away from the system. The lifetime of the metastable level E2 is about 5 msec.

Ruby laser has another absorption band which can be used for pumping, in the spectral range 350–450 nm. It is difficult to achieve continuous operation of a Ruby laser since it is a three level laser. However, in 1962, by using very intensive pump, using arc lamp with high pressure Mercury vapor, a continuous wave Ruby laser was build. The energy gap between the upper lasing level and the ground level in a Ruby laser is 1.789 eV. The wavelength of the emitted light out of a Ruby laser is 694.3 nm. The Schematic diagram of the First Ruby Laser is shown in Figure 2.6. The optical and laser properties of ruby at room temperature are given in Table 2.2.

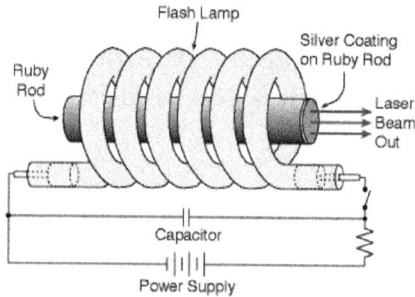

Figure 2.6   Schematic diagram of the First Ruby Laser

Table 2.2   Optical and laser properties of ruby at room temperature

Property	Values
Cr_3O_2 doping	0.05 Wt. %
Cr concentration	1.58×10^{19} ions cm^{-3}
Output wavelengths, 25°C	694.3 and 692.6 nm
Fluorescent lifetime (g)	3.0 ms at 300 K
Photon energy (hv)	2.86×10^{19} Ws
Quantum efficiency	0.7
Separation of R1 and R2 lines	29 cm
Major pump bands	Blue (4040), Green (5540)
Refractive index (n) at 694.3 nm	1.763 ordinary ray ; 1.755 extraordinary ray
Brewster angle	60° 37' at 694.3 nm
Maximum extractable energy	2.35 J/cm (complete inversion)

Property	Values
Maximum upper–state energy density	4.52 J/cm (complete inversion)
Upper–state energy at threshold	2.18/cm
Density	3.98 g/cc
Melting Point	2040°C
Young's Modulus	345 Gpa
Hardness	9 Mhos, 2000 Knoop
Birefringence	0.008

Nd DOPED MATERIALS From the large number of Nd:doped materials, only few have prominence. Nd:YAG, laser is the most important solid-state laser for scientific, medical, industrial and military application due to its high gain, good thermal and mechanical properties. Nd:glass is important for laser fusion drivers because it can be produced in large sizes. Very recently, Nd:Cr:GSGG has received considerable attention because of the good spectral match between the flashlamp emission and the absorption of the Cr ions. An efficient energy transfer between the Cr and Nd ions results in a highly efficient Nd:laser. Nd:YLF is a good candidate for certain specialized applications because the output is polarized and the crystal exhibits lower thermal birefringence. Nd:YLF has a higher energy storage capability due to its lower gain coefficient compared to Nd:YAG. Further, its output wavelength matches with that of phosphate Nd:glass. Hence mode-locked and Q-switched Nd:YLF lasers have become the standard oscillators for large glass lasers employed in fusion research.

Nd:YAG The Nd:YAG laser is the most commonly used solid-state laser. Neodymium-doped yttrium aluminium garnet (Nd:YAG) possesses a combination of properties uniquely favourable for laser operation. The YAG host is hard and possesses good optical quality and a high thermal conductivity. Furthermore, the cubic structure of YAG favours a narrow fluorescent line-width which results in high gain and low threshold for laser operation. In Nd:YAG, trivalent neodymium substitutes for trivalent yttrium, so charge compensation is not required. By improving the quality of the material and pumping techniques CW power can be obtained to several hundred watts from a single laser rod. On the other hand, in single-crystal Nd:YAG fiber lasers, threshold was achieved with absorbed pump powers as small as 1 mW. Today, more than thirty years after its first operation, the Nd:YAG laser has emerged as the most versatile solid-state system. In addition to the very favourable spectral and lasing characteristics displayed by Nd:YAG, the host lattice is noteworthy for its unique physical, chemical and mechanical properties. The Nd:YAG laser and its energy level diagram are shown in Figure 2.7(a) and 2.7(b), respectively.

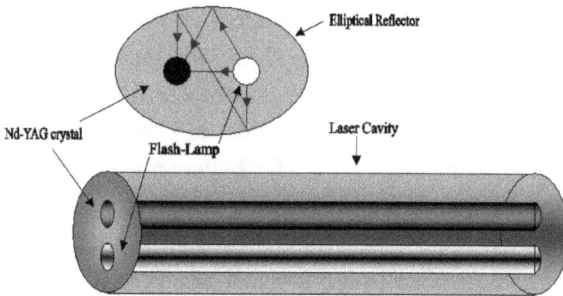

Figure 2.7   (a) Nd:YAG laser

The YAG structure is stable from the lowest temperature to the melting point and no transformations have been reported in the solid phase. The strength and hardness of YAG are lower than ruby but still high enough, so that normal fabrication procedures do not produce any serious breakage problems. Pure $Y_3Al_5O_{12}$ is a colourless, optically isotropic crystal which possesses a cubic structure characteristic of garnets. In Nd:YAG, about 1% of Y^{3+} is substituted by Nd^{3+}. The radii of the two rare ions differ by 3%. Therefore, with the addition of large amounts of neodymium, strained crystals are obtained, indicating that either the solubility limit of neodymium is exceeded or that the lattice of YAG is seriously distorted by the inclusion of neodymium.

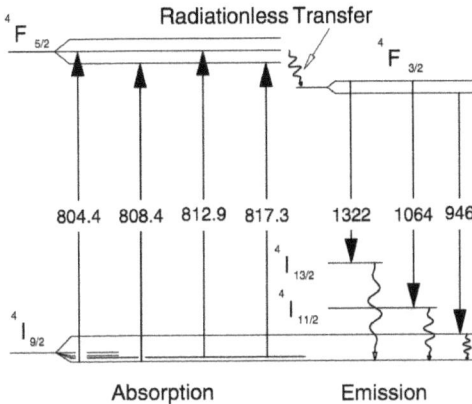

Figure 2.7 (b) Energy level diagram of Nd.YAG laser

Commercially available laser crystals are grown exclusively by the Czochralski method. The high manufacturing costs of Nd:YAG are mainly caused by the very slow growth rate of Nd:YAG which is of the order of 0.5 mm/h. Typical boules of 10 to 15 cm in length require a growth run of several weeks. However, all Nd:YAG crystals grown by Czochralski techniques show a bright core running along the length of the crystals. The cores originate from the presence of facets on the growth interface which have a different distribution coefficient for neodymium than the surrounding growth surface. That is, in order to provide rods of a given diameter, the crystal must be grown with a diameter that is somewhat more than twice as large. The boules are processed by quartering into sections. At present times, rods can be fabricated with maximum diameters of about 10 mm and lengths up to 150 mm. The optical quality of such rods is normally quite good and comparable to the best quality of Czochralski ruby or optical glass. For example, 6 mm by 100 mm rods cut from the outer sections of 20 mm by 150 mm boules typically may show only 1 to 2 fringes in a Twyman-Green interferometer.

Table 2.3   Physical and optical properties of Nd:YAG

Properties	Values
Chemical Formula	$Nd:Y_3Al_5O_{12}$
Crystal Structure	Cubic
Lattice Constants	12.01
Concentration	$\sim 1.2 \times 1020\ cm^{-3}$
Melting Point	1970 °C
Density	$4.56\ g/cm^3$
Mohs Hardness	8.5 Mohs
Refractive Index	1.82
Thermal Expansion Coefficient	7.8×10^{-6} /K [111], 0 – 250°C
Thermal Conductivity	14 W/m /K @ 20°C, 10.5 W /m/K @ 100°C.
Lasing Wavelength	1064 nm
Stimulated Emission Cross Section	$2.8 \times 10^{-19}\ cm^{-2}$
Relaxation Time of Terminal Lasing Level	30 ns
Radiative Lifetime	550 ms
Spontaneous Fluorescence	230 ms
Loss Coefficient	$0.003\ cm^{-1}$ @ 1064 nm
Effective Emission Cross Section	$2.8 \times 10^{-19}\ cm^2$
Pump Wavelength	807.5 nm
Absorption band at pump wavelength	1 nm
Linewidth	0.6 nm
Polarized Emission	Unpolarized
Thermal Birefringence	High

Neodymium concentration in YAG has been limited to 0.1 to 1.5%. Higher doping levels tend to shorten the fluorescent lifetime, broaden the line-width, and cause strain in the crystal, resulting in poor optical quality. In specifying Nd:YAG rods, the emphasis is on size, dimensional tolerance, doping level and passive optical tests of rod quality. In a particular application the performance of a Nd:YAG laser can be somewhat improved by the choice of the optimum Nd concentration. As a general guideline, it can be said that a high doping concentration (approximately 1.2%) is desirable for Q-switch operation because this will lead to high energy storage. For CW operation, a low doping concentration (0.6 to 0.8%) is usually chosen to obtain beam quality. It is worth noting that in contrast to a liquid or a glass, a crystal host is not amenable to uniform dopant concentration. This problem arises as a result of the crystal growth mechanism. In the substitution of the larger Nd^{3+} for a Y^{3+} in $Y_3Al_5O_{12}$, the neodymium is preferentially retained in the melt. The increase in concentration of Nd from the seed to the final stage of a 20 cm long boule is about 20 to 25%. For a laser rod 3 to 8 cm long, this end-to-end variation may be 0.05 to 0.10% of Nd_2O_3 by weight. The physical and optical properties of Nd:YAG are given in Table 2.3.

Nd:Glass There are several characteristics which distinguish glass from other solid-state laser host materials. Its properties are isotropic. It can be doped at very high concentrations with excellent uniformity and it can be made to large pieces of diffraction-limited optical quality. In addition, glass lasers have been made, in a variety of shapes and sizes, from fibers, rods and discs. There is wide variety of Nd-doped laser glasses depending on the compositions of the glass network and the ions. Among various laser glasses, only silicates and phosphates are commercially available at present with sufficient optical, mechanical and chemical properties.

There are two important differences between glass and crystal lasers. First, the thermal conductivity of glass is considerably lower than that of most crystal hosts. Secondly, the emission and absorption lines of ions in glasses are inherently broader than in crystals. A wider line increases the laser threshold value of amplification. Nevertheless, this broadening has an advantage. A broader line offers the possibility of obtaining and amplifying shorter light pulses and in addition, it permits the storage of larger amounts of energy in the amplifying medium for the same linear amplification coefficient. Thus, glass and crystalline lasers complement to each other. For continuous or very high repetition rate operation, crystalline lasers provide higher gain and greater thermal conductivity. Glasses are more suitable for high-energy pulsed operation because of their large size, flexibility in their physical parameters and the broadened fluorescent line. Unlike many crystals, the concentration of the active ions can be very high in glass. The practical limit is determined from the fact that the fluorescent lifetime and therefore the efficiency of stimulated emission, decreases with higher concentrations. In silicate glass, this decrease becomes noticeable at a concentration of 5% Nd_2O_3. The main constituents of glasses are non-metal oxides, such as SiO_2, B_2O_3 and P_2O_5. Different metal oxides alter the structure in various ways and make it possible to obtain a large variety of properties. The components are mixed before melting with the laser activators. The mixture is heated in a heat-resistant crucible. The principal laser glass manufactures use either platinum or ceramic crucibles, clay pots or ceramic continuous tanks to contain the melt. When the melt has reached a high viscosity, it is cast into a mold. Finally, the glass in the mold is placed into an annealing furnace where it is very slowly cooled Nd doped phosphate glass have been widely used in high average power laser solid state lasers, laser material processing, range finder and other industrial and scientific applications.

Glass laser rods are fabricated in a large variety of sizes. Typical rod sizes are between 10 and 50 cm in length, with diameters from 1 to 3 cm. However, rods up to 1 m in length and 10 cm in diameter are commercially available. Standard rod end configurations are the same as those mentioned for ruby and Nd:YAG rods. Besides Nd:YAG and Nd:glass, a lot of Nd-doped laser crystals have been doped. Some important ones are Nd:YLiF$_4$ (YLF) for low thermal lensing, Nd:YAlO$_4$ (YALO or YAP) for plolarised output, Cr,Nd:Gd$_3$ (Sc,Ga)$_5$O$_{12}$ (GSGG) for high efficiency and Cr,Nd:Y$_3$(Sc,Ga)$_5$O$_{12}$ (YSGG) for high efficiency.

Nd,Cr:GSGG Soon after the invention of the Nd:YAG laser, attempts were made to increase the efficiency of transferring radiation from the pump source to the laser crystal using a second dopant called a "sensitizer". A particularly attractive sensitizer is Cr^{3+} because the broad absorption bands of chromium can efficiently absorb light throughout the whole visible region of the spectrum.

The higher pump efficiency of Nd,Cr:GSGG does not automatically translate into better system performance because Nd,Cr:GSGG does exhibit much stronger thermal focusing and stress birefringence, compared to Nd:YAG. The absorption efficiency and the heat-deposition rate for the Nd,Cr:GSGG rod is almost three times those of Nd:YAG. Hence thermal focusing power as a function of lamp input power is several times larger in Nd,Cr:GSGG than in Nd:YAG. Therefore, if beam brightness is the criteria, rather than output energy, some of the advantage of GSGG is offset, particularly at high average powers. Tunability of the emission in solid-state lasers is achieved when the stimulated emission of photons is intimately coupled to the emission of vibrational quanta (phonons) in a crystal lattice. In these "vibronic" lasers, the total energy of the lasing transition is fixed but can be partitioned between photons and phonons in a continuous fashion. This results in broad wavelength tunability of the laser output. In other words, the existence of tunable solid-state lasers is due to the subtle interplay between the Coulomb field of the lasing ion, the crystals field of the host lattice, and electron-phonon coupling permitting broad-band absorption and emission. Therefore, the gain in vibronic lasers depends on transitions between coupled vibrational and electronic states. That is, a phonon is either emitted or absorbed with each electronic transition. The best known and most used vibronic lasers now available is the Alexandrite and Ti:sapphire lasers.

ALEXANDRITE   Alexandrite is a colour changing variety of the mineral chrysoberyl. The chemical composition of chrysoberyl is $BeAl_2O_4$ and it occurs in granitic pegmatite and mica schist. This is a rare oxide mineral that has two very rare varieties. The first is the colour changing alexandrite and the second is the cat's eye chrysoberyl. Alexandrite is the best-characterized commercially developed vibronic laser. Alexandrite is the common name for chromium-doped chrysoberyl, with four units of $BeAl_2O_4$ forming an orthorhombic structure. The crystal is grown in large boules by

Czochralski method much like ruby and YAG. The main problem is handling the very toxic beryllium Oxide. In any case, laser rods up to 1cm in diameter and 10 cm long with a nominal 2-fringe total optical distortion are commercially available. The chromium concentration of alexandrite is expressed in terms of the percentage of aluminium ions in the crystal which have been replaced by chromium ions. The Cr^{3+} dopant concentration, occupying the Al^{3+} sites, can be as high as 0.4 atomic percent and still yield crystals of good optical quality. In Alexandrite, concentration of 0.1 atomic percent represents 3.51×10^{10} Chromium ions per cubic centimeter. Alexandrite shows the typical absorption spectrum of Cr doped materials and is shown in Figure 2.8.

Absorption spectrum of Alexandrite has chromium lines in the red (doublet plus two other lines), a broad band in the yellow-green and two narrow bands in the blue. The most dramatic feature of alexandrite is the colour change. This effect is caused by combination of factors. The presence of the chromium in combination with the fact that alexandrite is doubly refractive and bi-axial allows alexandrite to possess three different refractive indexes in its three different optical directions. Each of these has a strongly different absorption spectrum, causing different colour to be seen. Daylight contains high proportions of blue light, so the stones appear green. Incandescent light contains a higher portion of red light, so the stones appear red. Many variances of the shades of the colours seen occur because each alexandrite can absorb light of different wavelengths many different ways.

It is optically and mechanically similar to ruby and possesses many of the physical and chemical properties of a good laser host. Alexandrite possesses hardness, strength, chemical stability and high thermal conductivity equal to two-thirds that of ruby and twice that of YAG. This enables alexandrite rods to be pumped at high average powers without thermal fracture. Alexandrite has a thermal fracture limit which is 60% that of ruby and five times higher than YAG. Surface damage tests using focused 750 nm radiation indicate that alexandrite is at least as resistant as ruby.

Alexandrite absorption, 0.176% Cr, E // b-axis

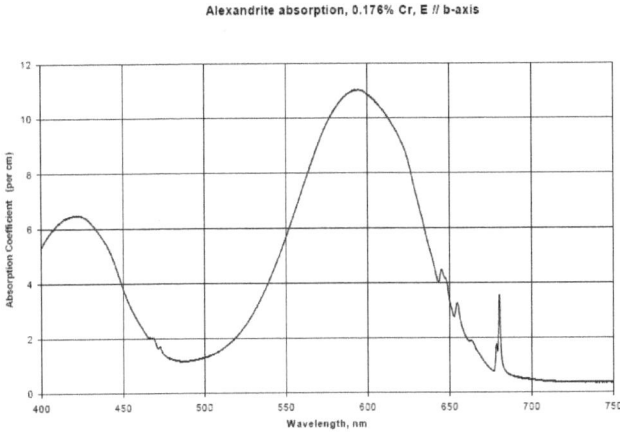

Figure 2.8 Absorption spectrum of Alexandrite. (A strong doublet at 680.5 and 678.5 nm and weak lines at 665, 655 and 645 nm)

The development of the alexandrite laser has reached maturity after nearly 10 years of efforts. Its current high average-power performance is 100 W, if operated at 100 Hz. Overall efficiency is close to 0.5%. Tunability over the range of approximately 700 to 818 nm has been demonstrated with tuning accomplished in a manner similar to dye lasers – a combination of etalons and birefringement filters. With these standard spectral control devices, 0.5 cm line widths and tunability over 150nm has been achieved. Alexandrite has been lased in pulsed and CW modes. It can also be Q-switched and mode-locked.

When a rod 10 cm long and 0.63 cm in diameter, is Q-switched in a stable resonator, it gives long-pulses (5J), with pulse duration less than 30 ns. The reason for such high output energies is that alexandrite is a low-gain medium (g = 0.04 cm – 0.1 cm) at room temperature.

Commercially available alexandrite lasers feature continuous, automatic tuning and a minimum 100mJ of Q-switched output with 0.3 nm band-width over the wavelength range from 730–780 nm. Because of alexandrine's physical strength and thermal properties, CW operation is possible at room temperature. The bulk of the experimentation to date has been with CW xenon arc lamps. CW operation with arc lamp pumping has proved difficult to achieve, yet output powers of up to 40W with good transverse-mode quality are now being generated. CW lasers were also acousto-optically Q-switched at rates greater than 10 KHz, with peak powers as high as 300W and pulse width of 1 fs. The physical and optical properties of alexandrite are listed in Table 2.4.

Table 2.4 Physical and optical properties of alexandrite

Properties	Values
Chemical composition	$BeAl_2O_4$
Crystal system	Orthorhombic
Habit	Tabular or prismatic; also trillings (repeated twinning producing pseudo–hexagonal or cyclic crystals)
Cleavage	Distinct (prismatic)
Fracture	Conchoidal to uneven
Hardness	8.5 Mohs
Toughness	8
Specific gravity	3.72
Lustre	Vitreous
Refractive index	1.746–1.755
Double refraction and optic sign	0.008 – 0.010, positive
Dispersion	low (0.015)
Pleochroism	Strong in alexandrite (green,yellowish, pink – in daylight; red, orange yellow, green – in filament light).
Luminescence	Red fluorescence in long wave UV and Weak red in short wave UV

Ti:Sapphire The $Ti:Al_2O_3$ laser is one of the more promising tunable solid-state systems, combining a broad tuning range

(800 nm peak, 300 nm band-width) with a relatively large gain cross-section (50% of Nd:YAG`s value at its peak). One of the greatest advantages is the material properties of the sapphire host itself, namely very high thermal conductivity, exceptional chemical inertness and mechanical rigidity. Titanium sapphire is presently available from commercial vendors in sizes of 3.5 cm diameter by 15 cm long optical quality, due to the well-developed growth technology for sapphire. The absorption and emission spectra for Ti:Al$_2$O$_3$ are shown in Figure 2.9. The broad, widely separated absorption and fluorescence bands are caused by the strong coupling between the ion and host lattice, and they are the key to broadly tunable laser operation.

Development of Ti:Al$_2$O$_3$ laser has been confined mainly to optical pumping due to short spontaneous emission life-time (3.2 µs) of the material. Flashlamp pumping of titanium sapphire gives a serious technological problem in terms of flashlamps lifetime. It was not possible earlier to get the short pulse duration needed for driving this crystals above threshold from the flashlamp. Only recently the development of special flashlamp allowed the construction of Ti:Al$_2$O$_3$ laser with high power and reasonable flashlamps lifetimes, in excess of 1 millions shots.

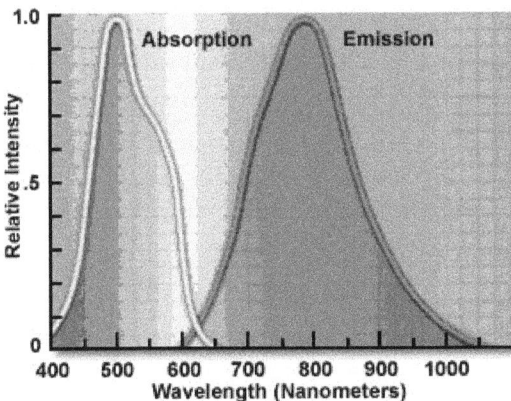

Figure 2.9 Absorption and emission Spectra of Ti:Sapphire

PUMPING   Many different light sources have been used for the optical excitation of solid state lasers. They can be divided into two broad classes viz., the standard incoherent sources and the coherent sources. The incoherent sources are used for pumping until now, because of their lower cost and easy availability. The maximum average power available from lamp pumped solid state lasers is also of order of magnitude higher than that available from laser pumped systems. This situation is now evolving, because of the recent availability of relatively high power semiconductors lasers, but their high cost limit the average power level in the watt range. On the other hand, lamp pumped systems are commercially available in the KW range. The incoherent light sources used for solid state lasers pumping are

1. **Noble-gas arc and flashlamps**   They are the most common sources, and comes in many different shapes and technology according to the application of the laser.

2. **Metal vapour arc lamps**   They are different from the previous lamps due to the narrow emission spectrum which sometimes leads to a higher efficiency. The main drawbacks of metal vapour arc lamps are the low power capabilities and the CW operation.

3. **Filament lamps**   They are made by the same tungsten halogen technology used for illumination and hence they are very cheap and the power supply is very simple. Due to their well-known drawbacks, such as the limited available average power, the low transfer efficiency at CW operation and the broad band blackbody emission spectrum, they are confined to low power and inexpensive Nd:YAG systems

4. **Sun**   There are several examples of solid state lasers pumped by the solar radiation, using mirrors to collect the light on the crystals. Of course, they depend on the availability of the sources but they are very interesting

for space application. In some cases, for example using a Nd,Cr:GSGG crystal, the overall efficiency compete with that of photovoltaic solar cells.

5. **Chemical reactions** They are mainly single shot devices because the chemicals used for the reaction must be replaced after each shot. The most common of these sources is the flash-bulb similar to that used in old photographic flash-lights.

6. **Light emitting diodes** LEDs are similar to diode lasers in concept but their broader emission bands and especially their wide emission angle give a lower coupling efficiency. The flashlamps are gas discharge devices generally filled with xenon or krypton gases which give pulsed radiation. But, the arc lamps are basically the same gas discharge devices but designed and optimized to produce continuous radiation. The most common shape of both flash and arc lamps is linear. Other shapes are also available from lamps manufactures. In the early stage of ruby lasers, the helical lamps, U-shaped and point lamps were employed. Linear lamps are now preferred for laser pumping since they are much easier to cool by means of a coaxial water flow confined by a transparent tube. In addition, a cylindrical emitter can be more easily coupled to a laser rod. Since almost 50% of the lamp input power is dissipated as heat, lamp body needs cooling. For small power systems, namely, low energy, low repetition rate pulsed lasers, static or forced gas cooling can be used. For higher power systems, liquid cooling can be used.

Although liquid cooling of flashlamps and D.C krypton arc lamps is occasionally performed using coolants other than water, deionized purified water continues to be most popular liquid coolant. The purity of the water must be high to prevent the etching of quartz and the attachment of dark deposits to the outside surface of the lamp envelop. In systems where electrical connections and coolant are in contact with one another, the water must be deionized to prevent the shorting or weakening of the starting pulse and to minimize erosion of electrical contacts due to electrolysis. Water resistivity greater than 0.2 mega ohms is highly recommended. All Cooling systems component materials should be plastic, stainless steel, or nickel plated metal. The flow velocity of the coolant around the flashlamp must be high in such a way the heat exchange between the lamp walls and the liquid is maximum.

PUMPING CAVITIES   Since, lamps and most other incoherent sources, emit light diffusely in any direction, an optical systems is used to collect the light into the laser active medium. This is usually done with the reflector surrounding the lamp and the laser crystals. The most widely used pump cavity is a highly reflective elliptical cylinder with the laser rod and pump lamp at each focus. The elliptical configurations are based on the geometrical theorem that rays originating from one focus of an ellipse are reflected into the other focus. Therefore, an elliptical configuration transfers energy from a linear source placed at one focus to a linear absorber placed along the second focus. The elliptical configuration is closed by two plane – parallel and highly reflecting end plates. This makes the elliptical cylinder optically infinite long.

A single elliptical cylinder can have a cross section with a large or small eccentricity. In one case, the laser rod and lamp are separated by a fairly large distance while in the second case they are close together. If the elliptical cylinder closely surrounds the lamp and rod, then the configuration is known as close-coupled elliptical geometry. This geometry usually results in the most efficient cavity. This cavity has further advantage that it minimizes the weight and size of the laser heads. Such reflectors are made by machining from solid metal blocks usually brass or aluminium. The reflective surface is then optical grade polished and coated with a suitable reflective film. For Nd:YAG or Nd:glass, the best laser coating is electroplated gold which has a good reflectivity on the main absorption bands of Nd. For crystals requiring blue or ultraviolet pumping (as ruby or Ti:Sapphire), vacuum evaporated aluminum is often used, since it gives higher reflectivity than gold at wavelengths shorter than 500 nm. The other possible technology is to close-coupled, nonfocussing, pump cavity, where the lamp and the rod are placed as close as possible and a reflector closely surrounds them. The various pumping configuration are given in Figure 2.10.

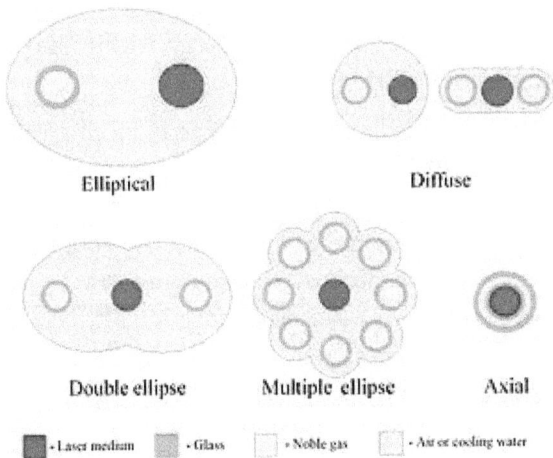

Elliptical   Diffuse

Double ellipse   Multiple ellipse   Axial

■ - Laser medium   - Glass   - Noble gas   - Air or cooling water

Figure 2.10   Various pumping configuration

Since all high average power solid state lasers are pumped by broad-band flash or arc lamps, thermal effects dominate in their operating characteristics. Typically, 5–10% of the power from the exciting flashlamps is converted to heat within the host material. This heating is due to:

1. Energy loss when ions decay from the pump bands to the upper laser level

2. Non-radiative decay of a fraction of the ions in the upper laser level.

3. Direct excitation of the host material by the broadband flashlamp light.

The heat is removed by cooling the laser rod. This leads to the development of a near parabolic temperature profile in the material. The temperature distribution takes between a few seconds (Nd:YAG) or a minute (Nd:glass) to stabilize. The resulting temperature gradients in an operating laser influence optical behaviour. The two mechanisms encountered for this purpose are

1. They lead to variations in refractive index causing lensing and wavefront distortion, and

2. They produce variations of stress within the body of the material inducing birefringence.

At very high flashlamp powers, the induced stress can lead to rupture of the laser rod. There are some possible techniques which can be used to compensate for thermal effects in laser rods. Several of these methods can be incorporated within a resonator design. Birefringence problems can be reduced in crystals by selecting the proper orientation of the crystals axes relative to the cylindrical axis. In commercially grown YAG, the rod is along the (111) direction. However, it has been shown that the (001) direction would be preferable for lower losses if the heat dissipation of the rod is lesser than 50 W. Generally, losses become independent of orientation axis for powers in excess of 50 W. Unfortunately, cutting the (001) direction rod from a (111) grown boule of Nd:YAG limits the size and material waste. Compensation of thermal focusing has been accomplished in many ways, for example, by grinding the ends of the rod to form a negative lens. The cavity mirror selection depends on the negative lens and the rod focal length in the resonator design. It should be noted that each of these passive schemes only provides compensation at one (selected) average pump power. However, many applications require a broad variation of the output power and more complicated adaptive resonator should be used (adaptive mirrors and phase conjugation, etc.).

Since heat is concentrated at the central axis of the cylindrical ruby rod in ruby laser, the material can be removed close to the central axis of the ruby rod. In this way, we can remove the heat concentration at the center of the rod. But this gives an annular laser output.

SEMICONDUCTOR LASER PUMPING The most important alternative to flashlamp pumping of solid–state lasers the diode laser. In the last twenty five years, numerous laboratory devices have been assembled which incorporate single diode lasers, small laser diode arrays or LEDs for pumping of Nd:YAG, Nd:glass and host of other Nd lasers. The low power output, low packaging density and extremely high cost of diode lasers prevented any serious applications for laser pumping in the past. The reason for the continued interest in this area stems from the potential dramatic increase in systems efficiency and component lifetime and reduction of thermal load of the solid-state laser material. The latter will not only reduce thermo-optic effects (which lead to better beam quality) but will also increase in pulse repetition frequency. The attractive operating parameters combined with low-voltage operation and the compactness of solid-state laser system have a high potential in applications.

The high pumping efficiency compared to flashlamps stems from the good spectral match between the laser-diode emission and the Nd absorption bands. Actually, flashlamps have higher radiation output to electrical input efficiency (70%) compared to laser diodes (25–50%).

The spectral match between the diode-laser emission and the long-wavelength Nd absorption band reduces the heat deposition in the laser material. Added to this, system lifetime and reliability will be higher in laser-diode-pumped solid state lasers compared to flashlamp based systems. Laser-diode arrays have exhibited lifetimes of the order of 19,000 hours in CW operation and 109 shots in the pulsed mode. Flashlamp life is on the order of 107 shots and about 200 hours for CW operation.

In addition, the high pump flux combined with a substantial UV content in lamp-pumped systems cause material degradation in the pump cavity and in the coolant which lead to systems degradation and contribute to maintenance requirements. Such problems are virtually eliminated with laser-diode pump sources. The absence of high-voltage pulses, high temperatures and UV radiation encountered with arc lamps are the special operating features of laser-diode-pumped systems.

In contrast to flashlamps or filament lamps, the emission from laser diode is highly directional and hence require different pump configurations. The techniques for transferring pump radiation from laser diode arrays to the solid-state laser material are discussed below.

We have to distinguish between optical system employed for end-pumping of a laser crystal, side-pumping of a cylindrical rod, and side-pumping of a rectangular slab. The focused end-pumping configuration is the most efficient pump-radiation transfer scheme, provided the pump light distribution and resonator mode matches (Figure 2.8). Since, the pump beam from the diode array is collinear with the optical resonator, the overlap between the pumped volume and the TEM_{00} mode can be very high. In addition, the coupling efficiency is high as long as the absorption length is equal to crystal length. In its basic form, end pumping involves a collimating lens with a large numerical aperture in order to collect radiation from a large cone angle and a focusing lens to produce a small diameter spot inside the laser crystal. Since the radiation emitted from a diode array has a high degree of astigmatism, one can introduce prisms or cylindrical optics to transform the beam into a circular shape. The purpose of the coupling optics is to shape the radiation distribution from the diode array such that the pumping volume coincides with the laser's TEM00 mode. Radiation from the laser diode array diverges approximately 40 degrees in the plane perpendicular to the active layer. The radiation is collected by a lens. After passing through a pair of prisms, the radiation is focused on the rod by a lens. The resulting divergence inside the Nd:YAG rod is about 6 degrees. In the most common geometry, a plano-concave configuration with high-reflection coating at 1.06 μm at one planar rod surface and anti reflection coating at 0.81 μm is chosen to allow the pump light to enter the rod. The intra cavity rod surface is anti reflection coated at 1.06μm and an output mirror can be placed either outside the rod or coated directly on the rod itself.

End pumping of a miniature Nd:YAG laser diode arrays is an attractive means of obtaining efficient CW lasers. However, at present, the end–pump scheme is useful only for low power lasers due to the smaller pump area. In order to achieve a somewhat higher output power, two sources can be polarizably coupled to double the pump power or a double-ended pumping arrangement can be employed.

In Side pumping configuration, the diode arrays are placed along the length of the laser rod and pumped perpendicularly to the direction of propagation of the laser. As more power is required, more diode arrays can be added along and around the laser rod. There are three practical approaches to couple the radiation emitted by the diode lasers to the rod:

a. direct coupling,
b. with optics between sources and absorber and
c. fiber optics coupling.

The direct coupling option does not permit for variations other than the position of the diode lasers around the rod. Fiber-optics coupling is very impractical for a large number of diode lasers. Optical coupling can be achieved by using optics such as lenses or elliptical and parabolic mirrors or by non-imaging optics such as reflective or refractive flux concentrators.

The side-pump geometry is not as efficient as the end-pump design. However, it permits scaling of the laser to large power levels.

As a matter of fact, the side-pump geometries are normally due to a small absorption length, low pumping density and wasted pump energy due to resonator-mode and pump-distribution mismatch. Large lasers require large rod diameters which increase the pump efficiency. They also require densely packed high-power diode arrays which provide gains over the whole cross section of the rod at sufficient intensity to permit the use of unstable resonators. All these factors contribute to a more efficient utilization of the pump radiation as compared to small systems.

The side-pump geometry is most suitable for slab laser pumping. Side-pumped slab lasers are usually pumped by 2-D arrays or densely-packed linear arrays. In a slab laser, the face of the crystal and the emitting surface of the laser diodes are in close proximity and no optics is employed. The slab can be pumped from one face only. In this case, the opposite face is bonded to a copper heat sink containing a reflective coating to get back the unused pump radiation to the slab for a second pass. An antireflection coating on the pump face is used to reduce coupling losses. Further, liquid cooling is employed to remove heat from the YAG and diode heat sink.

The laser diode pumped solid state lasers (DPSSL) are employed to perform the following applications summarized in Table 2.5.

The largest application of solid state lasers are in military equipment where they are used in range finders and target designators. Beyond that, solid state lasers are important in material processing, in medicine and in scientific research that are summarized in Table 2.6.

Table 2.5   Applications of diode-pumped solid-state lasers

Laser power	Application area	Example
< 10 W	Laboratory light source	Spectroscopy, Kinetics, CARS, interferometery,
	Semiconductor microfabrication	Link blowing, Thin film trimming
	Entertainment	Projection TV
	Communication	Communications space based point to point fiber laser repeaters
>10W	Materials Processing	Metals, Cutting and welding etc.,
	Semiconductor microfabrication	Link blowing, Thin film trimming
	communication	Space based point to Area: Drive for soft X–ray source
> kW	Fusion power	Laser reactor driver

Table 2.6 Solid State Lasers

Laser gain medium and type	Operation wavelength(s)	Pump source	Applications
Ruby solid-state laser	694.3 nm	Flashlamp	Holography, tattoo removal. The first type of laser invented, in 1960.
Neodymium YAG (Nd:YAG) solid-state laser	1.064 μm, (1.32 μm)	Flashlamp, laser diode	Material processing, range finding, laser target designation, surgery, research, pumping other lasers (in combination with frequency doubling). One of the most common high power lasers. Usually pulsed (down to fractions of a nanosecond)
Neodymium YLF (Nd:YLF) solid-state laser	1.047 and 1.053 μm	Flashlamp, laser diode	Mostly used for pulsed pumping of certain types of pulsed Ti:sapphire lasers, in combination with frequency doubling.

Laser	Wavelength	Pump source	Notes
Neodymium YVO4 (Nd:YVO) solid-state laser	1.064 μm	laser diode	Mostly used for continuous pumping of mode-locked Ti:Sapphire lasers, in combination with frequency doubling.
Neodymium Glass (Nd:Glass) solid-state laser	~1.062 μm (Silicate glasses), ~1.054 μm (Phosphate glasses)	Flashlamp, laser diode	Used in extremely high power (Terawatt scale), high energy (Mega joules) multiple beam systems for inertial confinement fusion. Nd:Glass lasers are usually frequency tripled to the third harmonic at 351 nm in laser fusion devices.
Titanium sapphire (Ti:sapphire) solid-state laser	650–1100 nm	Other laser	Spectroscopy, LIDAR, research. This material is often used in highly-tunable mode-locked infrared lasers to produce ultra-short pulses and in amplifier lasers to produce ultra-short and ultra-intense pulses.

Table 2.6 Continued

Laser gain medium and type	Operation wavelength(s)	Pump source	Applications
Thulium YAG (Tm:YAG) solid-state laser	2.0 μm	Laser diode	Laser radar
Ytterbium YAG (Yb:YAG) solid-state laser	1.03 μm	Laser diode, flashlamp	Optical refrigeration, materials processing, ultrashort pulse research, multiphoton microscopy, LIDAR.
Holmium YAG (Ho:YAG) solid-state laser	2.1 μm	Laser diode	Tissue ablation, kidney stone removal, dentistry.
Cerium doped lithium strontium(or calcium) aluminum fluoride (Ce:LiSAF, Ce:LiCAF)	~280 to 316 nm	Frequency quadrupled Nd:YAG laser pumped, excimer laser pumped, copper vapor laser pumped.	Remote atmospheric sensing, LIDAR, optics research.

(Contd.,)

Laser type	Wavelength	Pump source	Notes
Promethium 147 doped phosphate glass solid-state laser	933 nm, 1098 nm		Laser material is radioactive. Once demonstrated in use at LLNL in 1987, room temperature 4 level lasing in 147Pm doped into a lead–indium–phosphate glass étalon.
Chromium doped Chrysoberyl (Alexandrite) solid-state laser	Typically tuned in the range of 700 to 820 nm	Flashlamp, laser diode, mercury arc (for CW mode operation)	Dermatological uses, LIDAR, laser machining.
Erbium doped fiber laser	1.53–1.56 μm	Laser diode	Optical amplifier for telecommunications over optical fiber.
Uranium doped calcium fluoride (U:CaF2) solid-state laser	2.5 μm	Flashlamp	First 4-level solid-state laser (1960) developed by Peter Sorokin and Mirek Stevenson, second laser invented overall (after Maiman's ruby laser), liquid helium cooled, not in use today.

2.3.4 Semiconductors Lasers

At present, nearly 80 percent of the entire laser market is dominated by semiconductor lasers or laser arrays which is used to pump other solid-state lasers or operate with a coherent output from a two dimensional array. Output powers in excess of one kilowatt with more than 10 percent power efficiency should become commercially available. These lasers will provide a major boost for industrial processing and medical applications. Semiconductor lasers are widely used as source for optical communication.

Monolithic diode lasers which operate without an external cavity have spectral linewidths that are typically 30 MHz at output power levels of about 10 mW. Such spectral linewidths are not useful for a wide range of applications such as radar, optical heterodyne communications and high resolution spectroscopy.

The source of the spectral broadening in these semiconductor diode lasers is fundamental and quantum in nature and arises from the spontaneous emission noise within the laser. This spontaneous emission noise causes the phase of the electric field vector of the radiation to fluctuate which inturn leads to a spectral broadening. In addition, damped relaxation oscillators due to the perturbation of the field amplitude by spontaneous emission give rise to side bands. These side bands are typically 1–2 GHz away from the main peak. When such devices are placed into high Q external cavities, the linewidths are significantly reduced. The linewidths of less than 400 Hz have been produced in such lasers.

The frequency stability characteristics of free running external cavity GaAlAs diode lasers which have already been demonstrated are listed below:

Extending the useful frequency range of presently available diode lasers such as GaAlAs can be accomplished by the use of efficient non linear frequency conversion. Wide band gap semiconductor diode lasers have been very difficult to operate at wavelengths shorter than 600 nm. The materials of choice would be the II–VI compounds, but so far have not proven to operate as diodes because of the inability to fabricate p–type material. Therefore, nonlinear frequency conversion would be one solution for the near term operation of compact blue diode laser sources.

Property	Values
Fundamental width	< 1000 Hz
Future capability	< 10 MHz
Short term jitter	5 KHz / 30 ms
Long term jitter	7 KHz
Center frequency drift	5 KHz/h
Locking stability	< 10 mHz
Baseband noise	+/– 0.5 GHz < 100 dB

The blue laser is a laser that emits electromagnetic radiation at a wavelength of between 360 and 480 nm, which the human eye sees as blue light or else light at the blue end of the spectrum. Diode lasers which emit light at 445 nm are becoming popular as laser pointers. Lasers emitting wavelengths below 445 nm appear violet to the human eye, a distinctly different color. This is true, for example, of the most commercially–common "blue" lasers, the diode lasers used in Blu–ray applications, which emit 405 nm light that is violet. This light causes fluorescence in some chemicals, in the same way that ultraviolet or "black light" does. The class of blue lasers are frequently semiconductor laser diodes based on gallium(III) nitride (violet color) or indium gallium nitride (often true–blue in color, but able to produce other colors, as well). Both blue and violet lasers can also be constructed using frequency-doubling of infrared laser wavelengths from diode lasers or diode–pumped lasers. Devices that employ blue laser light have applications in many areas ranging from optoelectronic data storage at high density to medical applications.

Major advances in high power diode lasers have been made in the world with the development of the graded index separate confinement heterostructure GaAlAs diode laser. These lasers have produced over one watt of CW power at room temperature with more than 50% power conversion efficiency. One problem with these monolithic devices is the poor spectral output. This may be improved by operation in an external cavity. Because the broadening mechanism in semiconductor lasers leads to a homogeneous gain line, all of the laser output will occur in a single frequency when the devices operated in a single spatial mode. Such a diode laser has been operated in an external cavity and has produced more than 1.5 watts of continuous power at room temperature in a spectral width of less than 8 GHz and a spatial beam quality which was better than 2–3 times diffraction limited. Because of the energy band properties of quantum-well laser materials, the gain bandwidths can be very broad. A single such laser when operated in an external cavity with a grating has been continuously tuned from 790 to 860 nm at room temperature. Two such lasers have covered the range from 720 to 870 nm. Frequency doubling of such high power devices would produce useful radiation from 360 to 435 nm with additional tuning over the entire visible range possible by parametric conversion.

The use of efficient diode lasers and diode laser arrays are now in widespread use as efficient pump sources for solid-state lasers to provide all solid state, compact, reliable and efficient lasers which are able to produce output from a few mW to 100 W. Such lasers would find numerous applications where the disadvantages of lamp pumping are obvious. The most common material that has been pumped by diode lasers is Nd:YAG which has its main output lines at 1.06 and 1.32 μm.

Laser Diodes

Compound Semiconductors

In addition to the elemental semiconductors, Compound semiconductors were developed for several optoelectronic applications. They are

1. Binary compounds

   GaAs, AlAs, GaP, etc. and (III–V) compounds. III–V compounds widely used in optoelectronic and high-speed applications.

   ZnS, ZnTe, CdSe and (II–VI) compounds.

   SiC, SiGe (IV compounds). Si widely used for rectifiers, transistors, and ICs

2. Ternary compounds - GaAsP.

3. Quaternary compounds - InGaAsP.

A laser diode is a laser where the active medium is a semiconductor similar to that found in a light-emitting diode. The most common and practical type of laser diode is formed from a p–n junction and powered by injected electric current. These devices are sometimes referred to as injection laser diodes to distinguish them from (optically) pumped laser diodes.

A laser diode, like many other semiconductor devices, is formed by doping a very thin layer on the surface of a crystal wafer. The crystal is doped to produce an n-type region and a p-type region, one above the other, resulting in a p–n junction, or diode.

As charge injection is a distinguishing feature of diode lasers as compared to all other lasers, they are also known as injection lasers. When an electron and a hole are present in the same region, they may recombine or "annihilate" causing spontaneous emission which is necessary to initiate laser oscillation. In a conventional semiconductor junction diode, the energy released from the recombination of electrons and holes is carried away as phonons, i.e., lattice vibrations, rather than as photons.

The difference between the photon-emitting semiconductor laser (or LED) and conventional phonon-emitting (non–light-emitting) semiconductor junction diodes lies in their band gaps. The photon–emitting semiconductors are the direct bandgap semiconductors, while the phonon emitting semiconductors are the indirect band gap semiconductors. The silicon and germanium are direct band gap semiconductors. Other materials such as Gallium arsenide, indium phosphide, gallium antimonide and gallium nitride are the compound semiconductors possessing indirect band gap emit light. A simple laser diode is shown in Figure 2.11.

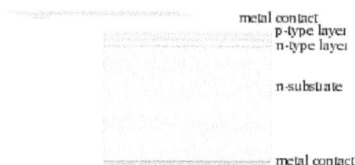

Figure 2.11 A simple laser diode

The spontaneous and stimulated emission processes are vastly more efficient in direct bandgap semiconductors than in indirect bandgap semiconductors and therefore silicon is not a common material for laser diodes.

As in other lasers, the gain region is surrounded with an optical cavity to form a laser. In the simplest form of laser diode, an optical waveguide is made on that crystal surface, such that the light is confined to a relatively narrow line. The two ends of the crystal are cleaved to form perfectly smooth, parallel edges, forming a Fabry–Perot resonator. Photons emitted into a mode of the waveguide travel along the waveguide and be reflected several times from each end face before they are emitted. As a light wave passes through the cavity, it is amplified by stimulated emission, but light is also lost due to absorption and by incomplete reflection from the end facets. Finally, if there is more amplification than loss, the diode begins to "lase".

Some important properties of laser diodes are determined by the geometry of the optical cavity. Generally, in the vertical direction, the light is contained in a very thin layer, and the structure supports only a single optical mode in the direction perpendicular to the layers. In the lateral direction, if the waveguide is wide compared to the wavelength of light, then the waveguide can support multiple lateral optical modes, and the laser is known as "multi-mode". These laterally multi-mode lasers are adequate in cases where one needs a very large amount of power, but not a small diffraction-limited beam. It is used in printing, activating chemicals, or pumping other types of lasers.

In applications where a small focused beam is needed, the waveguide must be made narrow, on the order of the optical wavelength. This way, only a single lateral mode is supported and one ends up with a diffraction-limited beam. Such single spatial mode devices are used for optical storage, laser pointers, and fiber optics. Note that these lasers may still support multiple longitudinal modes, and thus can lase at multiple wavelengths simultaneously.

The wavelength emitted is a function of the band-gap of the semiconductor and the modes of the optical cavity. In general, the maximum gain occur for photons with energy slightly above the band-gap energy and the modes nearest the gain peak lase most strongly. If the diode is driven strongly enough, additional side modes may also lase. Some laser diodes such as most visible lasers, operate at a single wavelength but that wavelength is unstable and changes due to fluctuations in current or temperature.

Due to diffraction, the beam diverges rapidly after leaving the chip, typically at 30 degrees vertically by 10 degrees laterally. A lens must be used in order to form a collimated beam like that produced by a laser pointer. If a circular beam is required, cylindrical lenses and other optics are used. For single spatial mode lasers, using symmetrical lenses, the collimated beam ends up being elliptical in shape, due to the difference in the vertical and lateral divergences. This is easily observable with a red laser pointer.

The simple diode described above is extremely inefficient and has been heavily modified in recent years to accommodate modern technology, resulting in a variety of types of laser diodes, as described below.

1. Double heterostructure lasers (DH Lasers)

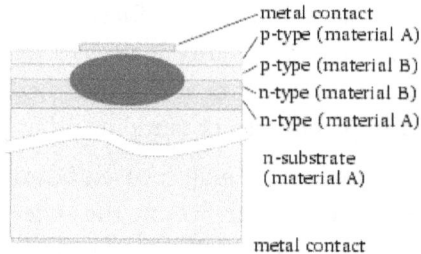

metal contact
p-type (material A)
p-type (material B)
n-type (material B)
n-type (material A)

n-substrate
(material A)

metal contact

Figure 2.12   Front view of a double heterostructure laser diode

The front view of a double heterostructure laser diode is shown in Figure 2.12. In these devices, a layer of low bandgap material is sandwiched between two high bandgap layers. One commonly-used pair of materials is gallium arsenide (GaAs) with aluminium gallium arsenide ($AlxGa(1–x)As$). Each of the junctions between different bandgap materials is called a heterostructure, hence the name "double heterostructure laser" or DH laser.

The advantage of a DH laser is that the region where free electrons and holes exist simultaneously (the active region) is confined to the thin middle layer. This means that many more of the electron-hole pairs can contribute to amplification – not so many is left out in the poorly amplifying periphery. In addition, light is reflected from the hetero junction and hence, the light is confined to the region where the amplification takes place.

2. Vertical-cavity surface-emitting laser (VCSELs)

A simple VCSEL structure is shown in Figure 2.13. Vertical-cavity surface-emitting lasers (VCSELs) have the optical cavity axis along the direction of current flow rather than perpendicular to the current flow as in conventional laser diodes. The active region length is very short compared with the lateral dimensions so that the radiation emerges from the surface of the cavity rather than from its edge. The reflectors at the ends of the cavity are dielectric mirrors made from alternating high and low refractive index quarter–wave thick multilayer.

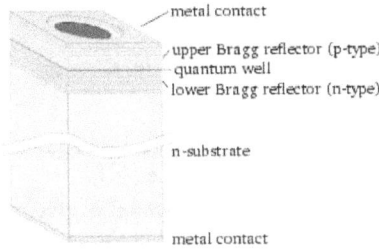

metal contact
upper Bragg reflector (p-type)
quantum well
lower Bragg reflector (n-type)
n-substrate
metal contact

Figure 2.13 A simple VCSEL structure

Such dielectric mirrors provide a high degree of wavelength-selective reflectance at the required free surface wavelength λ if the thicknesses of alternating layers d_1 and d_2 with refractive indices n_1 and n_2 are such that $n_1 d_1 + n_2 d_2 = 1/2\lambda$ which then leads to the constructive interference of all partially reflected waves at the interfaces. But there is a disadvantage: because of the high mirror reflectivities, VCSELs have lower output powers when compared to edge-emitting lasers.

There are several advantages to producing VCSELs when compared with the production process of edge-emitting lasers. Edge-emitters cannot be tested until the end of the production process. If the edge-emitter does not work, whether due to bad contacts or poor material growth quality, the production time and the processing materials have been wasted. Additionally, because VCSELs emit the beam perpendicular to the active region of the laser as opposed to parallel as with an edge emitter, tens of thousands of VCSELs can be processed simultaneously on a three-inch Gallium Arsenide wafer. Furthermore, even though the VCSEL production process is more labor and material intensive, the yield can be controlled to a more predictable outcome.

$Al_x GaAs_1 - x$ is the exciting source (lasers or LEDs) used in optical communication at 0.85 µm and they are termed as short wavelength sources. There is much interest in InGaAsP sources operating near 1.3 µm and 1.55 µm and they are called as long wavelength sources. The long wavelength sources have two advantages: The loss in optical fiber is lower than 0.85 µm and the dispersion is nearly zero.

The lasers and the commonly used components for optical communications are given below:

Components	Short wavelength–1st window (0.85 µm)	Long wavelength–2nd window (1.3 µm)	Long wavelength–3nd window (1.55 µm)
Laser	AlGaAs	InGaAsP	InGaAsP
Fiber loss	2.8 dB/km	0.5 dB/km	0.2 dB/km
Detector	Si	Ge or InGaAs	InGaAs

The properties of semiconductor laser diodes used in optical communication as sources are given below:

Type	Wavelength (Nm)	Power (mW)	Current (mA)	Spectral width (Nm)	Beam divergence (Degrees)
AlGaAs –LED	830	1	200	40	30 × 42
AlGaAs–Double heterojunction laser	850	10	300	2.5	5 × 42
AlGaAs –laser	830	15	65	0.1	13 × 40
InGaAsP–laser	1300, 1550	7	250	4	10 × 30
Multiple – stripe AlGaAs laser	850	500	1600	2	10 × 35

The applications of semiconductor laser diodes are given below:

Semiconductor lasers

Laser gain medium and type	Operation wavelength(s)	Pump source	Applications
Semiconductor laser diode	wavelength depends on device material: 0.4 μm (GaN) or 0.63–1.55 μm (AlGaAs) or 3–20 μm (lead salt)	Electrical current	Telecommunications, holography, laser pointers, printing, pump sources for other lasers. The 780 nm AlGaAs laser diode, used in compact disc players, is the most common type of laser in the world.

The different structures of diode lasers are illustrated in Table 2.7.

Table 2.7   Different Structures of Diode Lasers

Properties	Values
Chemical Formula	$Nd:Y_3Al_5O_{12}$
Crystal Structure	Cubic
Lattice Constants	12.01
Concentration	~ 1.2×1020 cm^{-3}
Melting Point	1970 °C
Density	4.56 g/cm^3
Mohs Hardness	8.5 Mohs
Refractive Index	1.82
Thermal Expansion Coefficient	7.8×10^{-6} /K [111], 0 – 250°C
Thermal Conductivity	14 W/m /K @ 20°C, 10.5 W /m/K @ 100°C.
Lasing Wavelength	1064 nm
Stimulated Emission Cross Section	2.8×10^{-19} cm^{-2}
Relaxation Time of Terminal Lasing Level	30 ns
Radiative Lifetime	550 ms
Spontaneous Fluorescence	230 ms
Loss Coefficient	0.003 cm^{-1} @ 1064 nm
Effective Emission Cross Section	2.8×10^{-19} cm^2
Pump Wavelength	807.5 nm
Absorption band at pump wavelength	1 nm
Linewidth	0.6 nm
Polarized Emission	Unpolarized

The other types of lasers such as gas lasers, ion lasers, metal vapour lasers and dye lasers are not discussed here as it is outside the scope of this book

2.4 PHOTON DETECTORS

Optical detectors are the components that convert the light energy of fiber optic communications into electrical signals (voltage or current) for recovery of data.

The optical detector allows the incident optical power (photon) to illuminate semiconductor device, resulting in the generation of hole–electron pairs. These electron–hole pairs produce an electric current in the presence of a electric field and detector responds to the current with a speed of tens of picoseconds. This is basic principle of operation of a *p–n*, *p–i–n* and avalanche photodiodes.

Photovoltaic detectors also known as photodiode contain a *p–n* semiconductor junction. It is a self generated detector when light strikes on it. The photovoltaic detector may operate without external bias voltage. The solar cell used on spacecraft and satellites is a very good example. For Al_xGaAs_{1-x} sources, silicon photo diodes are suitable detectors. These detectors offer excellent high frequency response at wavelengths 850 nm to 1.1 μm. They have peak spectral response near 0.9 μm which is very close to the wavelength of Al_xGaAs_{1-x} lasers. At longer wavelengths (1.3 μm), silicon photodiodes are not useful. Germanium or InGaAsSb photodiodes are suitable for longer wavelengths

2.4.1 Detector Characteristics

The performance of a detector is measured by the efficiency with which it converts optical power to electric current and by its speed of response.

1.   Responsivity is defined as the detector output per unit of input power. The units of responsivity is amperes/

watt. The responsivity gives no information about noise characteristics but the knowledge of the responsivity allows the user to determine how much detector signal will be available for a specific application.

2. Noise equivalent power (NEP) is defined as the radiant power that produces a signal voltage (current) equal to the noise voltage (current) of the detector. Since the noise depend on the bandwidth, it must be specified. The equation for NEP is

$$NEP = IAV_N / V_S(\Delta f)^{1/2}$$

Where, I is the irradiance incident on the detector area A, V_N is the root mean square noise voltage within the bandwidth Δ_f and V_S is the root mean square signal voltage. From the definition, it is apparent that lower the value of NEP, the better are the characteristics of the detector for detecting a small signal in the presence of noise. The NEP of a detector is dependent on the area of the detector.

3. Quantum efficiency of the detector is the measure of fraction of incident photons on the detector which produce electron-hole pairs which is less than or equal to unity. For a $p–n$ junction device, the quantum efficiency is the number of hole-electron pairs divided by the number of incident photons. For a $p–i–n$ diode the quantum efficiency can be less than unity.

4. Another important detector characteristic is the speed of the detector response to changes in light intensity.

5. Since photo detectors often are used for detection of fast pulses, a more important term called rise time is often used to describe the speed of the detector response. Rise time is defined as the time difference between the points at which the detector has reached 10% of its peak output and the point at which it has reached 90% of its peak

response, when it is irradiated by a very short pulse of light. The fall time is the time between the 90% point and the 10% point on the trailing edge of the pulse waveform. This is also called the decay time.

6. Another important characteristic of detectors is their linearity. Photo detectors are characterized by a response that is linear with incident intensity over a broad range.

All optical detectors respond to the power in the optical beam which is proportional to the square of the electric field. They are thus called "square-law detectors." All detectors for optical communication use optical absorption in a depletion region to convert photons into electron–hole pairs and then sense the number of pairs. The electron-hole pairs give rise to photocurrent (I_p) due to the electric field in the depletion region. The responsivity (R) of the detector is defined as the ratio between photocurrent and optical power (Pin).

$$R = I_p / P_{in} = 2_p h_Q q / hw = 2_p h_Q q / hn$$

where, h_Q is quantum efficiency and q is the charge generated per photon.

h_Q = electron–hole pair generation rate / photon incidence rate.

Detectors for optical communication are of three types; PN detector, PIN detector and APD.

Silicon, germanium and InGaAs semiconductors are used as detectors for optical communication. Their bandgaps, emission wavelengths and responsivity are given below.

The *p–n* photodiodes, *p–i–n* photodiodes and Avalanche photodiodes are used as common detectors in optical communication. For optical communication, LEDs and lasers have very high reliability, compact and mechanically stable.

Material	Bandgap (eV)	Wavelength (nm)	Wavelength of Peak response (nm)	Responsivity (Max) (A/W)
Si	1.17	300–1100	800	0.5
Ge	0.775	500–1800	1550	0.7
InGaAs	0.75–1.24	1000–1700	1700	1.1

2.4.2 P–N Photodiodes

A p–n junction can be used to detect light if reverse bias voltage is applied to the device. The most frequently used photodiode is silicon. Silicon photodiodes are widely used as the detector elements in optical disks and as the receiver elements in optical-fiber telecommunication systems operating at wavelengths around 800 nm. Silicon photodiodes respond over the approximate spectral range of 400–1100 nm, covering the visible and part of the near-infrared regions. Hence, they represent the most widely used type of laser detectors for lasers operating in these regions. Silicon photodiodes have become the detector of choice for many laser applications. The responsivity reaches a peak value around 0.7 amp/watt near 900 nm and decreases at longer and shorter wavelengths. Silicon photodiodes are useful for detection of many of the most common laser wavelengths including argon, He–Ne, AlGaAs and Nd:YAG.

2.4.3 PIN Detector Diodes

To increase the frequency response of photodiodes, PIN photodiode has been developed. Group III–V semiconductor material such a InGaAs/InP or InGaAs/InAlAs are used to fabricate p–i–n diodes. The doping, thickness and other properties are controlled to obtain the optical response, sensitivity, noise and other desired performance properties. The p–i–n and avalanche diodes are fabricated by forming the layers of p, i, n and n type material by epitaxial growth. The PIN construction features a layer of *p* type material, a layer of intrinsic silicon and a layer of *n* type materials.

A basic *p–n* diode with depletion layer under reverse bias is modified slightly to include an intrinsic region between the p and n regions to enlarge the depletion region sufficiently so that it becomes relatively insensitive to reverse bias level and hence will not change dimension under moderate illumination. Illumination of the intrinsic region will ionise valence electrons thereby creating a number of free electrons in the conduction band and holes in the valence band. Due to the strong field, the electrons and holes will be swept out of the junction region. This leads to current flow in the external circuit and temporarily lowers the device bias. This is picked up as an ac variation on the load resistor RL. The *p–i–n* structure can be a high speed structure with low power.

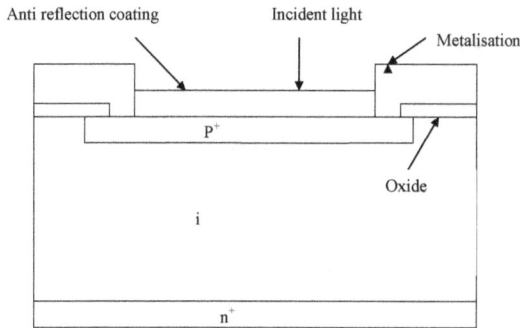

Figure 2. 14 Schematic diagram of a silicon p–i–n detector

Figure 2.14 shows a typical silicon *p–i–n* detector fabricated by diffusing dopanats through a series of masks to form the *p* and *n* regions in an i type substrate. This device has a layer of nearly intrinsic material bounded on one side by a relatively thin layer of highly doped *p*–type semiconductor and on the other side by a relatively thick layer of *n*–type semiconductor. A sufficiently large reverse bias voltage is applied so that the depletion layer, from which free carriers are swept out, spreads to occupy the entire volume of intrinsic material. This volume then has a high and nearly constant electric field. It is called the depletion region because all mobile charges have been removed. Light that is absorbed in the intrinsic region produces free electron–hole pairs, provided that the photon energy is high enough. These carriers are swept across the region with high velocity and are collected in the heavily doped regions. The frequency response of such PIN photodiodes can be very high, of the order of 1010 Hz. This is higher than the frequency response of p–n junctions without the intrinsic region. The oxidising layer provides insulation. A thin layered antireflection coating is applied to reduce the reflection from the surface of the device which increases its quantum efficiency.

2.4.4 Avalanche Photodetectors (APD)

Basically, an APD is a p–n junction photodiode operated with high reverse bias. The materials is typically InP/InGaAs. The operating principle of APD is similar to that of the p–i–n diode but it has a much longer propagation path for the electron. The high applied potential and impact ionization from the light wave generates electron–hole pairs. Subsequently, they cause an avalanche across the potential barrier. This leads to an effective propagation gain. This current gain gives the APD, its greater sensitivity. APDs are commonly used up to 2.5 Gaps and sometimes to 10 Gaps. Avalanche photodetectors (APDs) are used in long-haul fiber optic systems since they have superior sensitivity.

The avalanche photodiode offers the possibility of internal gain and it is sometimes referred to as a solid-state photomultiplier.

Figure 2.15 depicts a silicon avalanche photodetector fabricated by diffusion processes. The high field region is formed by consecutive diffusion of p and n dopants. Uniformity and parameter control are critical in the fabrication of useful

APDs. Silicon avalanche photodiodes (APDs) are used in lower frequency systems (up to 1.5 or 2 GHz) where they meet the modest frequency response requirements and low cost.

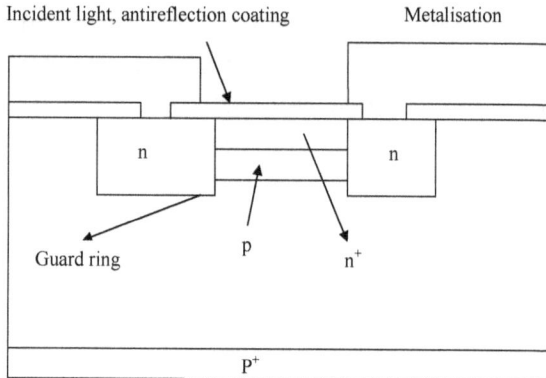

Figure 2.15 Schematic diagram of a silicon Avalanche photodetector

Higher output signals can be achieved by an avalanche diode. It uses a small internal current to generate a large output signals. It has the advantages of a good output at low light levels with wide dynamic range. That is, it can handle high and low light levels. However, the disadvantages of APD are

1. it has higher noise levels,

2. cost more,

3. generally requires higher operating voltages (in kilovolts) and

4. its gain decreases with an increase in temperature.

An APD has good electrical output in low light conditions. APD offers better performance (higher responsivity) at higher cost in comparison with PIN photodiode.

2.5 FIBER OPTIC COMMUNICATION SYSTEM

The block diagram of fiber optic communication system is shown in Figure 2.16. The electrical signal from input transducer (information to be transmitted) are modulated by a process called pulse code modulation in a coder. This produces a stream of equivalent digital electrical pulses which are changed by optical transmitter into light pulses for transmission by the optical fiber. The transmitter is either on light emitting diode or a laser diode. The infrared radiation is used for this purpose because it undergoes least scattering according to the formula for scattering power (p×a×1/λ4). At the receiving end, the optical receiver is a photo detector which converts the incoming infrared signals into the corresponding electrical impulses before they are processed by decoder. This is passed to the output transducer.

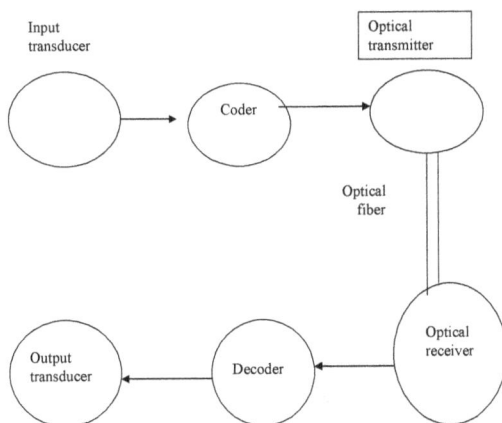

Figure 2.16 Block diagram of fiber optic communication system

The advantages of this optical fiber system are large. Some of them are,

1. It has high information carrying capacity – at present 30,000 telephone calls at once on a pair of fibers where as copper cable carry just 2000 calls only.

2. It is free from noise due to electrical interference.

3. Greater distances can be covered without amplifiers.

4. A cable of optical fiber is lighter, smaller and easier to handle than a copper cable.

5. Cross talk between adjacent channels is ruled out.

6. It offers greater security to the user.

2.6 SPLICERS AND CONNECTORS

2.6.1 Fiber Optic Splicers

A low loss connection between two fibers can be done by aligning the cores. The amount of light which is not coupled depends on the mechanical accuracy of the alignment, the physical match between the fiber parameters and the other materials in the gap between the fibers. The lateral offset, longitudinal offset and angular offset are a few alignment defects.

A splicing is a device to connect one fiber optic cable to another permanently. This technique provides lower insertion loss compared to fiber connection. Fiber optic cables have to be spliced together to realize a link of a particular length. Splicing is required at building entrances, wiring closets, couplers and intermediate point between transmitter and receiver. There are two basic types for splicing. They are fusion splicing and mechanical splicing.

Fusion splicing uses an electric arc to weld two fiber optic cables together. The splices done with computer controlled alignment of fiber optic cables offer low insertion loss (0.01 dB). This method provides lowest loss connection. The equipment fusion splicer performs optical splicing in two ways. a) It precisely aligns the two fibers and b) Generate a small electrical arc to melt the fibers and weld them together. With this equipment it is possible to achieve insertion loss less than 0.1 dB for both single mode and multimode fibers. On the other hand, fiber mechanical splicing share all common elements. They are easily applied in the field and require little or no tooling. In this splicing, the insertion loss is less than 0.5 dB.

The procedure for fiber optic cable splicing is as follows:

1. Strip fiber cable jacket

2. Stripe fiber tubes

3. Clean the cable gel

4. secure the cable tubes

5. Strip first splicing fiber

6. Place the fusion splice protection sleeve

7. Clean the bare fiber

8. Cleave the fiber to a specified length as per fusion splicer's manual

9. Prepare the second fiber being spliced

10. Place both fibers in the fusion splicer and do fusion splicing

2.6.2 Fiber Optic Connectors

If the connection between the fibers is rearrangeable as a demountable connection, they are called as connectors. Since 1980, several manufacturers developed many fiber optic connectors. Unlike electronic connectors, fiber optic connectors do not have jack and plug. In fiber optic connectors, a fiber mating sleeve (adapter or coupler) sits between two connectors. At the centre of the adapter, there is a cylindrical sleeve made by ceramic (zirconia) or phosphor bronze. Ferrules slide into the sleeve and mate to each other. The adapter body provides mechanism to hold the connector bodies such as snap–in, push and latch, twist on or screwed on. The following are the familiar fiber optic connectors.

1. **ST Connector** It is a single fiber cable (Simplex) which is easy to install. It is used as twist on mechanism and it is available in single mode and multimode.

2. **FC Connector** It is a single fiber cable (Simplex) which is easy to install. It is used as screw on mechanism and it is available in single mode and multimode.

3. **SC Connector** It is a single fiber cable (Simplex) and multi fiber connector (duplex) design. It is used as snap-in mechanism and it is available in single mode and multimode.

4. **FDDI connector** Duplex only, multimode only.

5. **LC Connector** Simplex and duplex, push and latch mechanism, 1.25 mm ceramic (zirconia) ferrule, available in single and multimode.

6. **MU Connector** simplex, duplex, snap in mechanism, 1.25 mm ferrule.

7. **E2000(LX.5) Connector** 1.25 mm ferrule, snap in mechanism.

8. **MT–RJ connector** duplex only, multimode only.

9. **VF–45 Connector** duplex only, no ferrules at all, Plug and jack version.

10. **Opti–Jack connector** duplex only, 2.5 mm ferrule, plug–jack version.

11. **Ribbon fiber connector** available in single and multimode.

12. **SMA 905 and SMA 906 connector** simplex only, multimode only.

13. **Biconic connector** simplex only, available in single and multi mode.

14. **D4 connector** 2.5 mm ferrule, screw on mechanism, simplex only.

REVIEW QUESTIONS

1. Draw the structure of LED and explain its working.

2. Distinguish between direct and indirect band gap semiconductors.

3. Why silicon is not used to construct photonic semiconductor?

4. Why Ohms law is not obeyed by LEDs?

5. Distinguish between spontaneous and stimulated emission.

6. What are Einstein's coefficients? Explain.

7. What is the importance of diode pumped solid state lasers?

8. Explain in detail the development of DPSSL.

9. Write note on non-linear optic crystals. What is second harmonic generation.

10. With necessary energy level diagram, explain the working of a ruby laser.

11. With necessary energy level diagram, explain the working of a Nd:YAG laser.

12. Write a note on (i) Alexandrite laser, (ii) Ti:Sapphire laser

13. Discuss on the various types pumping configuration.

14. Explain the salient features of a semiconductor laser.

15. What are compound semiconductors? Describe the construction and working of a laser diode.

16. Write a note on (a) Double hetero structure laser and (b) vertical cavity surface emitting injection laser.

17. What are detector characteristics? Explain.

18. Discuss the structure of P-i-N detector diode and explain its working.

19. Discuss the structure of Avalanche photo detector and explain its working.

    What is its importance?

20. With necessary block diagram, explain a fiber optic communication system

21. Distinguish between splicers and connectors.

Chapter III

OPTOELECTRONIC MODULATORS AND OPTICAL SENSORS

3.1 ELECTROPTIC EFFECT AND ELECTROPTIC MODULATOR

The linear electroptic (or electro-optic) effect, also known as Pockel effect is the change in the index of refraction along one or more axes in certain crystals and is proportional to the magnitude of an externally applied electric field (birefringence). Therefore, by applying a voltage across the electrodes of an electroptic crystal, one can change the phase of light as it passes through the crystal. By placing the crystal between crossed polarizers, this phase modulation can be converted into amplitude modulation. The Pockels effect occurs only in crystals that lack inversion symmetry, such as lithium niobate or gallium arsenide. An electroptic modulator is a device that uses an applied electrical field to alter the polarization properties of light.

The change in refractive index is dependent on the direction and polarization of the incident beam. The effect of electric field on the index of refraction polarizes the optical beam in an arbitrary direction in a crystal which is described by a third-rank tensor. The operation and application of electro-optic devices depend on birefringence which is induced by application of voltage to a crystal. Many crystalline materials exhibit birefringence naturally, without application of any voltage. The birefringence is present all the time. Examples of such crystals are quartz and calcite. There are also a number of crystals that are not birefringence naturally, but the voltage induces birefringence in them. This phenomenon is called the electroptic effect.

The electroptic modulator (EOM) is an optical device used to modulate a beam of light. The modulation may be imposed on the phase, frequency, amplitude or direction of the modulated beam. Modulation bandwidths extending into the gigahertz range are possible with the use of laser-controlled modulators. Generally, a nonlinear optical material (organic polymers) with an incident static or low frequency optical field experience modulation of its refractive index.

Electroptic and acoustoptic modulators are used with lasers as Q-switches, beam deflectors, light beam modulators and mode-lockers. An electroptic modulator (EOM) is a device which can be used for controlling the power, phase or polarization of a laser beam with an electrical signal. Electroptic modulators contain one or two Pockel cells along with polarizers and work on the principle of linear electro-optic effect. The non-linear crystal materials are used for fabricating electroptic modulators. They are potassium di-deuterium phosphate (DKDP), potassium titanyl phosphate (KTP), β-barium borate (BBO), lithium niobate ($LiNbO_3$), lithium tantalate ($LiTaO_3$), barium sodium niobate and ammonium dihydrogen phosphate (ADP). Added to these some inorganic and polymer materials are also used as electro-optic materials. *Optical switches* are modulators where the transmission is switched either on or off. Such a switch can be used for selecting certain pulses from a train of ultra short pulses or in cavity-dumped lasers and regenerative amplifiers. Electroptic modulators and acoustoptic modulators are used to mode lock lasers. The large apertures, outstanding modulation efficiency and selected crystals make EOMs as the right choice even for high-power lasers from UV to IR wavelengths. The essential requirements for efficient electroptic modulation are low half-wave (switching) voltage and broad 3-dB modulation bandwidth.

3.1.1 Electro-optic Modulators Types

Phase Modulators The simplest type of electro-optic modulator is a phase modulator containing only a Pockels cell where an applied electric field to the crystal changes the phase delay of a laser beam sent through the crystal. The polarization of the input beam often has to be aligned with one of the optical axes of the crystal so that the state of polarization is not changed. Many applications require a small periodic or nonperiodic phase modulation. The optical resonator in lasers uses electroptic phase modulators for monitoring and stabilizing a resonance frequency. Phase modulators are used to stabilize the frequency of a laser beam or to mode lock a laser. Phase modulators should not be used in interferometric applications. The electroptic crystals in phase modulators contain refractive index inhomogeneities.

Polarization Modulators Depending on the type and orientation of the nonlinear crystal and on the direction of the applied electric field, the phase delay depends on the polarization direction. A Pockels cell can thus be seen as a voltage-controlled waveplate, and it can be used for modulating the polarization state.

Amplitude Modulators When optical elements are combined with polarizers, Pockels cells can be used for other kinds of modulation. In particular, an amplitude modulator is based on a Pockels cell for modifying the state of polarization.

Electroptic amplitude and phase modulators allow us to control the amplitude, phase and polarization state of an optical beam electrically.

3.1.2 Electroptic Device as Light Beam Modulator

The use of an electroptic device as a light-beam modulator is shown in Figure 3.1. Polarized light is incident on the modulator.

The analyzer is oriented at 90° to the polarizer and it prevents any transmitted light when there is no voltage in the electro-optic material. When correct voltage is applied to the device, the direction of the polarization is rotated by 90°. Then the light l passes through the analyzer.

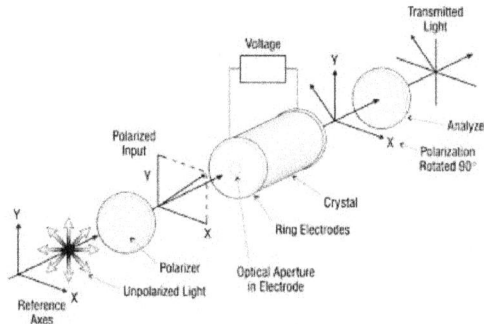

Figure 3.1 Electroptic device as a light beam modulator

There are two types of electro-optic effects. The first one is the Kerr electroptic effect seen in liquids such as nitrobenzene, while the second one is Pockels electroptic effect seen in crystalline materials such as ammonium dihydrogen phosphate or lithium niobate. The electro-optic modulators are also often called Pockels cells.

Electroptic modulators can be designed in different ways, either by applying voltage parallel to the light propagation or by using central aperture electrodes or transparent electrodes. This is called a longitudinal electroptic modulator. On the other hand, metal electrodes are on the sides of the crystal and the voltage is perpendicular to the light propagation. This is called a transverse electroptic modulator. Longitudinal and transverse modulators have different applications. The electroptic materials and some of their characteristics are shown in Table 3.1. Most of the materials are suited for use in the visible and near-infrared region of the electromagnetic spectrum, but cadmium telluride is useful as a modulator for CO_2 lasers in the long-wavelength infrared. The speed of electro-optic modulators is expressed as a bandwidth.

Table 3.1 Electro-optic materials and some of their characteristics

Materials	Transmission range (μm)	Band width (MHz)	Refractive index n_0, n_e at wavelength (μm)
Ammonium dihydrogen phosphate - ADP $NH_4H_2PO_4$	0.3–1.2	to 500	1.51, 1.47 at 1.06
Potassium dihydrogen phosphate – KDP – KH_2PO_4	0.25–1.7	> 100	1.51, 1.47 at 0.55
Potassium dideuterium phosphate – KDP KD_2PO_4	0.3–1.1	to 350	1.49, 1.46 at 1.06
Lithium niobate – LN– $LiNbO_3$	0.5–2	to 8000	2.23, 2.16 at 1.06
Lithium tantalite – $LiTaO_3$	0.4–1.1	to 1000	2.14, 2.143 at 1.00
Cadmium telluride – CdTe	2 – 16	to 1000	$n_0 = 2.6$ at 10

The advantage of electroptic modulators is its high speed and larger bandwidth than mechanical and acoustoptic devices. On the other hand, its disadvantage is due to the high voltage requirements and limited beam-diameter capabilities.

3.1.3 Applications

Some typical applications of electro-optic modulators are

a. modulating the power of a laser beam, e.g., in laser printing, high speed digital data recording or high speed optical communications

b. in laser frequency stabilization schemes,

c. Q switching of solid state lasers

d. Active mode locking

e. Switching pulses in pulse pickers, regenerative amplifiers and cavity dumped lasers

3.2 ACOUSTOPTIC DEVICES

The acoustoptic effect is a specific case of photoelasticity where there is a change of a material's permittivity due to a mechanical strain. The acoustoptic effect involves an interaction between sound waves and light waves traveling through typical crystals. The acoustoptic effect can be used to control the frequency, intensity and direction of an optical beam. Acoustoptic modulators are based on the diffraction of light by a column of sound in a suitable interaction medium. Acoustoptic modulator is a device that varies the amplitude and phase of a light beam (from a laser or by sound waves). It is also known as a Bragg cell. An acoustoptic device requires a material with good acoustic and optical properties and high optical transmission. Light diffracted by an acoustic wave of a single frequency produces two distinct diffraction types. These are Raman–Nath diffraction and Bragg diffraction. Raman–Nath diffraction is observed with relatively low acoustic frequencies, typically less than 10 MHz and with a small acousto-optic interaction length which is typically less than 1 cm. This type of diffraction occurs at an arbitrary angle of incidence. In contrast, Bragg diffraction occurs at higher acoustic frequencies usually exceeding 100 MHz.

There are three categories of acousto-optic devices such as acoustoptic modulator, filter and deflector.

A typical acoustoptic device is shown in Figure 3.2. In acoustoptic device, the piezoelectric materials and metal layers are bonded with acousto-optic material. Using metal layers as electrodes, radio-frequency is applied across the piezoelectric material. Then piezoelectric material generates acoustic wave in the acousto-optic medium. The propagating acoustic waves into the crystal form plane wavefronts. The opposite ends of the material from the transducer suppress reflected acoustic waves.

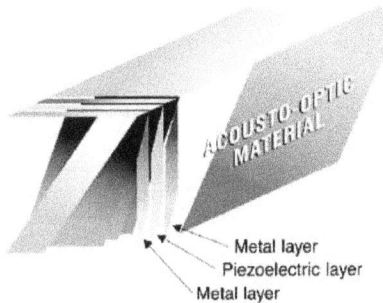

Metal layer
Piezoelectric layer
Metal layer

Figure 3.2 A typical acoustoptic device

The elastoptic properties of the medium produce a periodic variation of the index of refraction.

3.2.1 Acoustoptic Modulator

In acoustoptic modulator, the parameters of the acoustic wave such as the amplitude, phase, frequency and polarization are varied to modulate the optical properties. The acoustoptic interaction also makes it possible to modulate the optical beam by both temporal and spatial modulation.

A simple method of modulating the optical beam traveling through the acoustoptic device is done by switching the acoustic field on and off, and hence the acoustoptic device modulates the output along the Bragg diffraction angle. The device is operated as a modulator by keeping the acoustic wavelength (frequency) fixed and varying the drive power to vary the amount of light in the deflected beam.

Acoustoptic device as a modulator is shown in Figure 3.3. The deflected beam will give higher values of the extinction ratio than the undeflected beam. When the acoustic drive is in off position, the light in the direction of the deflected beam is zero. When the acoustic drive is in on position, light is diffracted into that direction. Hence, the acoustoptic device controls the light at on and off positions. The device is operated as a modulator by keeping the acoustic wavelength fixed and varying the drive power to vary the amount of light in the deflected beam.

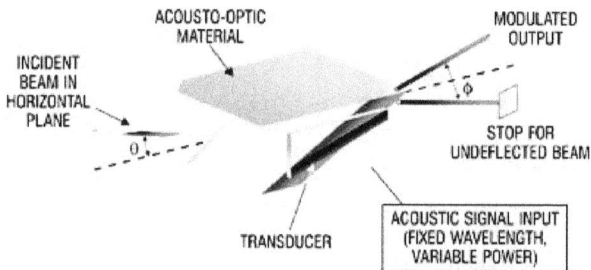

Figure 3.3 Acoustoptic device as a light-beam modulator

Acoustoptic light-beam modulators have a number of important desirable features such as low electric power and high extinction ratios. Acoustoptic devices are compact for systems where size and weight are important. As compared to electro-optic modulators, they have lower bandwidth but do not require high voltage.

The commercially available acoustoptic modulators and their properties are shown in Table 3.2.

Table 3.2 Acousto-optic modulators and their properties

Material	Spectral range (mm)	Figure of merit $(10^{-15}\,m^2/W)$	Bandwidth (MHz)	Typical drive power (W)	Refractive index	Acoustic velocity (m/sec)
Fused silica/ quartz – SiO_2	0.3 – 1.5	1.6	to 20	6	1.46 (6343 nm)	5900
Gallium arsenide–GaAs	1.0 – 11	104	to 350	1	3.37 (1.15 µm)	5340
Gallium phosphide–GaP	0.59 – 1.0	45	to 1000	50	3.31 (1.15 µm)	6320
Germanium–Ge	2.5 – 15	840	to 5	50	4.0 (10.6 µm)	5500
Lead molybdate–$PbMoO_4$	0.4 – 1.2	50	to 50	1 – 2	2.26 (633 nm)	3630
Tellurium dioxide–TeO_2	0.4 – 5	35	to 300	1 – 2	2.26 (633 nm)	4200
Lithium niobate–$LiNbO_3$	0.5–2	7	> 300	50–100	2.20 (633 nm)	6570

The acoustoptic modulators are used in

a.   Q switching of solid state lasers

b.   Cavity dumping of solid state lasers, generating either nanosecond or ultra short pulses

c.   Active mode locking

d.   Modulating the power of the laser beam in laser printing

e.   Shifting the frequency of a laser beam for various measurements.

3.2.2 Acousto-Optic Deflectors

These are essentially the same as acoustoptic modulators. In an acoustoptic modulator, only the amplitude of the sound wave is modulated whereas in acoustoptic deflectors both the amplitude and frequency are adjusted. An acoustoptic deflector spatially controls the optical beam. In an acoustoptic deflector, the power to the acoustic transducer is kept constant while the acoustic frequency is varied to deflect the beam to different angular positions. The acoustoptic deflector utilises the change in the angle as a function of the change in frequency.

3.2.3 Light-Beam Deflectors

These are sometimes called scanners, and are used for applications such as optical processing display and optical data storage. The acoustoptic device as a light-beam deflector is shown in Figure 3.4. Here, the drive power is kept constant and the acoustic wavelength is varied to deflect the beam to different angular positions.

The advantages of acousto-optic deflectors are

a.   easy control of deflection modes and position,

b.   simple structure,

c.   wide variety of uses from a single deflector,

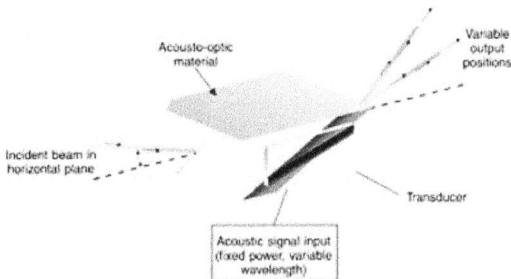

Figure 3.4 Acousto-optic device as a light-beam deflector

d. larger number of resolvable spots than electro-optic devices and

e. faster access than mechanical devices. Acoustoptic deflectors are used in applications such as high-frequency scanning and optical signal processing.

3.3 MAGNETOPTIC DEVICES

Magnetoptic effects are intimately connected to the magnetic properties of materials. When light is transmitted through a magnetoptic material, Faraday effect occurs. Faraday effect or Faraday rotation is a magnetoptical phenomenon or an interaction between light and a magnetic field. The rotation of the plane of polarization is proportional to the intensity of the component of the magnetic field in the direction of the beam of light. Magnetoptic Kerr effect (MOKE) is one of the magnetoptic effects. It describes changes of the reflections from the magnetized media. It is similar to the Faraday effect that describes the light passing through the media. The light that is reflected from a magnetized surface, changes in polarization. The magneto-optic effect breaks time reversal symmetry locally and Lorentz reciprocity. This is the necessary condition to fabricate optical isolator devices.

Magnetoptic materials have unique physical properties such as linear magnetoptic effect and nonreciprocal optical effects in dielectric media. This helps to construct devices with many special functions, which are not possible by photonic devices. The properties of these devices are polarization control, optical isolation, optical modulation, and magneto-optic recording. Magnetoptic activity is usually very small (less than 1^0) in conventional systems.

3.4 FIBER MODULATORS

The ferroelectric material lithium niobate ($LiNbO_3$) has been extensively applied to optical devices. It has excellent electroptic properties such as a large electro-optic effect and high-speed response. It is also transparent for infrared light and it is easy to fabricate into low-loss channel waveguides by diffusing titanium. $LiNbO_3$ devices are used in external intensity modulators, phase modulators, multi/demultiplexers and as switch arrays for optical fiber network systems. $LiNbO_3$ devices are also very useful for optical wavelength division multiplexing (WDM) systems because of the possibility of operation in the range of wide-wavelength infrared light with a single device due to its transparence. $LiNbO_3$ external modulators have been developed for extensive use in high speed and long-distance optical fiber transmission systems.

The integrated optical components for optical communications are generally fabricated using electroptic single crystal materials such as $LiNbO_3$. Among the integrated optical components, the contribution from electroptic modulators using $LiNbO_3$ waveguide structures has been significant in the last several decades due to their high-speed and chirp-free nature.

The $LiNbO_3$-based fiber modulators are suitable for applications with a high optical bandwidth and high data transfer rates in optical networks. They are used in optical communications, sensor technology and instruments, optical power meters, optical attenuator modules, defense and research.

3.5 INTERFEROMETRIC SENSORS

In the last few decades, telecommunication industries employ optical fibers widely due to their exemplary performance as the best light guidance. Further, optical fibers have been intensively investigated at various sensor fields owing to their unique characteristics such as multiplexing, remote sensing, high flexibility, low propagating loss, high sensitivity, low fabrication cost, small form factor, high accuracy, simultaneous sensing ability and immunity to electromagnetic interference. Fiber optic interferometer sensors for measuring temperature, strain, pressure, rotation, displacement, refractive index, polarization, ultrasound and so on are widely reported in the literature. Added to this, their sensing abilities have been considerably enhanced by utilizing innovative fiber optic technologies of fiber gratings, fiber interferometers, Brillouin/Raman scattering, surface Plasmon resonance (SPR), micro-structured fibers, nano-wires and specialty fiber couplers, *etc.* Indeed, some fiber optic sensors are used for real time deformation monitoring of aircrafts, ships, bridges and constructions. With the development of human-friendly smart materials, even health monitoring systems using fiber devices have attracted a great interest as future technologies. The recent review article by *Byeong Ha Lee et al* (Sensors 2012, 12, 2467–2486) gives a complete picture on this topic. The interferometric sensors are categorized as a) Fabry–Perot, b) Mach–Zehnder, c) Michelson and Sagnac. Specific examples of recently reported interferometric sensor technologies and its applications are also given under respective interferometric sensors.

A fiber optic interferometer uses the interference between two beams propagated through different optical paths of a single fiber or two different fibers. As expected, one of the optical paths is affected by external disturbances. Since the interferometers give a lot of temporal and spectral information as their signal, the changes in the wavelength, phase, intensity, frequency and bandwidth can be quantitatively determined with high accuracy and high sensitivity. Due to the fiber optic sensors, small-sized fiber devices became reality for measuring various physical parameters with high accuracy.

3.5.1 Fabry–Perot Interferometer Sensor

The Fabry–Perot interferometer (FPI) is an optical instrument which uses multiple-beam interference. A Fabry–Perot interferometer (Fabry–Perot etalon) is a linear optical resonator which consists of two highly reflecting mirrors (with some small transmittivity) and is often used as a high-resolution optical spectrometer. One exploits the fact that the transmission through such a resonator exhibits sharp resonances. Interference occurs due to the multiple super positions of both reflected and transmitted beams at two parallel surfaces. For the fiber optic cases, the FPI can be simply formed by intentionally building up reflectors inside or outside of fibers. FPI sensors can be largely classified into two categories as extrinsic and intrinsic. The extrinsic FPI sensor uses the reflections from an external cavity formed out of the interesting fiber. Although the extrinsic FPI is easy to fabricate with low cost, it has disadvantages of low coupling efficiency, careful alignment and packaging problem. On the other hand, the intrinsic FPI fiber sensors have reflecting components within the fiber itself. When disturbance is introduced to the sensor, the phase difference is influenced with the variation in the optical path length difference of the interferometer. Applying longitudinal strain to the FPI sensor changes the physical length of the cavity or/and the RI of the cavity material. This results in phase variation. By measuring the wavelength shift in spectrum of a FPI, the strain applied on it can be quantitatively obtained. Fiber optic sensors measure temperature, pressure, refractive index, strain, displacement, force and load using the Fabry–Perot principle. For obtaining the RI of liquid, extrinsic FPI sensors are appropriate. The schematic diagram of an extrinsic FPI liquid RI sensor system based on a PCF lens is shown in Figure 3.5.

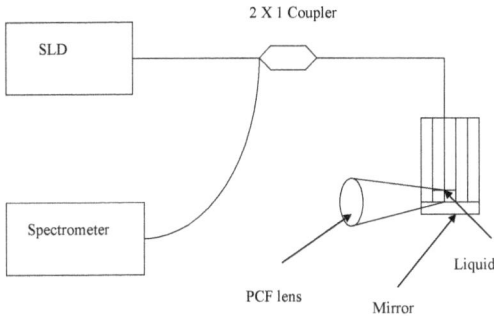

Figure 3.5 Schematic diagram of an extrinsic Fabry–Perot Interferometer liquid RI sensor system based on a PCF lens

Figure 3.5 illustrates the extrinsic sensor configuration based on a phonic crystal fiber (PCF) lens. The problem of low coupling efficiency of extrinsic FPI sensors could be overcome by introducing a PCF and a fiber lens on it. By using the electric arc discharge of a conventional fusion splicer, the air holes of the PCF could be collapsed and a lens was easily formed on its distal end. The PCF lens having a properly optimized curvature effectively acted as both a beam reflector and a collimator at the same time. The spectrum of a fabricated sensor is measured with an air cavity.

The extrinsic structure has the merit of sensing displacement since the phase value of FPI signal can affect directly by the displacement of the external reflecting surface. The pressure and ultrasound sensors are widely used with the extrinsic FPI configuration using polymer thin films as the reflecting surfaces. As the Young's modulus of polymer is much lower than that of fiber material, the polymer film can be used as a deformable cavity for the measurement of pressure or ultrasound.

In addition to the extrinsic FPI sensors, a variety of intrinsic FPI sensors have been developed with various fiber structures recently. Among them, a double cavity structure fiber FPI is unique and interesting. The double cavity fiber FPI can also be used as a temperature sensor.

Since the holey optical fiber (HOF) cavity and the multi mode fiber (MMF) cavity have different thermo-optic coefficients, the temperature-induced movement of the first peak is different from the movements of the other peaks. By coating a gas sensitive material on the end face of the MMF cavity, it can also be utilized as a gas sensor. Hydrogen gas sensor can also be realized by coating hydrogen-sensitive metal palladium. Similarly, multi cavity FPI biosensor has become reality by chemically etching the cores of single mode fibers (SMFs) and fusion-splicing them in series. In double cavity FPI fiber sensors, the inner cavity is not affected by the variation of the outer chemical and physical environments. Hence, power fluctuation or external disturbance is compensated by using the first Fourier peak.

3.5.2 Mach–Zehnder Interferometer Sensors

Mach–Zehnder interferometers (MZIs) are commonly used in diverse sensing applications due to their flexible configurations. In the early MZIs, a beam splits into two beams viz., the reference and the sensing beams. Subsequently, they recombine by using two fiber couplers. The recombined light has the interference component according to the OPD between the two beams. For sensing applications, the reference beam is kept isolated from external variation and only the sensing beam is exposed to the external variation. The variation in the sensing beam due to the external parameters such as temperature or strain or refractive index changes the OPD of the MZI. This can be easily detected by analyzing the variation in the interference signal. The schematic diagram of Mach–Zehnder Interferometer is shown in Figure 3.6

After the advent of long period fiber gratings (LPGs), the usage of two different beams in the MZIs were replaced with the scheme of in-line waveguide interferometer. In this set up, a part of the beam guided as the core mode of a SMF is coupled to cladding modes of the same fiber by an LPG and then re-coupled to the core mode by another LPG. The combined beam and the uncoupled beam in the core cause interference which gives a compact but very effective MZI. It has the same physical lengths in both the reference beam and the sensing beam but has the different optical path lengths due to the modal dispersion.

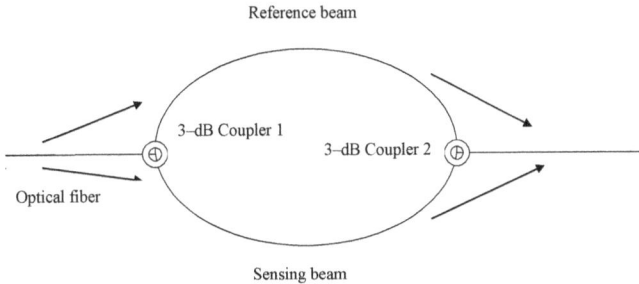

Reference beam

3–dB Coupler 1

3–dB Coupler 2

Optical fiber

Sensing beam

Figure 3.6   The schematic diagram of Mach–Zehnder Interferometer

An MZI temperature sensor using LPGs was reported by *Kim et al* [2002]. In that paper, they analysed thermo-optic coefficient of the fiber core material using a pair of LPGs. The fine interference fringe enabled to calculate even the wavelength dependency of the effective index of the core mode. It also showed that the Germanium-doped core had a stronger thermo-optic coefficient than the Boron co-doped core. An RI sensor based on the MZI composed of a pair of LPGs has also been reported by *Allsop et al* [2002].

However, the LPG pair MZI has a disadvantage in the operating wavelength. The LPG works only in a limited band(s) of wavelengths due to the phase matching phenomenon of fiber gratings. Added to this, both LPGs should be identical to get the maximum performance.

In recent times, simultaneous measurements of several parameters are possible with the in-line MZI. By taking the advantage of LPG pair made in double cladding fiber (DCF), one can simultaneously measure strain and temperature. Simultaneous sensor could be made by applying strain only at the grating-free region of an LPG pair while temperature was applied to the whole region of the LPG pair. With this configuration, the phase and the envelope of the MZI interference fringes were thermally shifted with the same rate but the strain shifted only the phase. Hence, it is possible to measure temperature and strain simultaneously.

3.5.3 Michelson Interferometric Sensor

Michelson interferometric sensors are used as position sensor, vibration sensor and refractive index sensor. It is used for the measurement of very small displacement. Fiber-optic sensors based on Michelson interferometers (MIs) are quite similar to MZIs. The basic concept is the interference between two beams which are reflected at the end of mirrors in an MI as shown in Figure 3.7. In fact, an MI is like a half of an MZI in configuration.

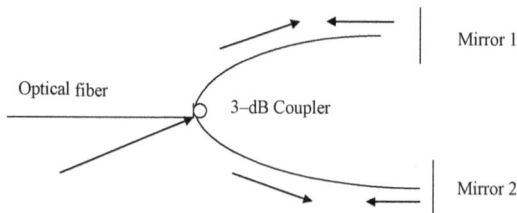

Figure 3.7 Basic configuration of a Michelson interferometer

Thus, the fabrication method and the operation principle of MIs are almost the same as MZIs. The main difference is the existence of reflectors. Since MIs use reflection modes, they are compact and handy in practical uses and installation. Multiplexing capability with parallel connection of several sensors is another beneficial point of MIs. However, it is essential to adjust the fiber length difference between the reference beam and the sensing beam of an MI within the coherence length of the light source.

An in-line configuration of Michelson interferometer is shown with Figure 3.8. A part of the core mode beam is coupled to the cladding mode(s), which is reflected along with the uncoupled core mode beam by the common reflector at the end of the fiber.

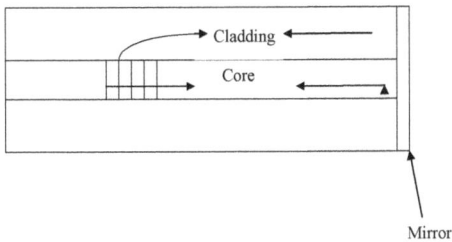

Mirror

Figure 3.8 Schematic diagram of a compact in-line Michelson interferometer

There are several fiber-optic sensors based on in-line MIs, especially for measurements of temperature and RIs of liquid specimens. The RI sensor based on a single LPG is easy to use practically in comparison with the MZI type LPG pair sensor because the grating-free region alone can be immersed in the liquid specimen.

Another important MI application is the measurement of flow velocity. To make the MI configuration appropriate to a flow velocity sensor, tapering method was used for splitting a beam into two cores of a twin-core fiber. Due to the induced flow, there appears a bending of the twin-core fiber. This sensor measures the optical path difference change between beams in the two cores.

3.5.4 Sagnac Interferometer Sensor

In recent years, Sagnac interferometers (SIs) are becoming popular for various sensing applications due to their simple structure, easy fabrication and environmental robustness. An SI consists of an optical fiber loop, along which two beams are propagating in counter directions with different polarization states. The schematic diagram of sensor based on Sagnac interferometer is shown in Figure 3.9

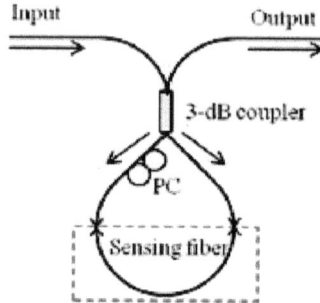

Figure 3.9 Schematic of the sensor based on a Sagnac interferometer (Courtesy: Byeong Ha Lee et al, 2012)

Here, the input light is split into two directions by a 3-dB fiber coupler and the two counter-propagating beams are combined again at the same coupler. Unlike other fiber optic interferometers, the OPD is determined by the polarization dependent propagating speed of the mode guided along the loop. To maximize the polarization-dependent feature of SIs, birefringent fibers are typically utilized in sensing parts. The polarizations are adjusted by a polarization controller (PC) attached at the beginning of the sensing fiber. The signal at the output port of the fiber coupler is governed by the interference between the beams polarized along the slow axis and the fast axis. The phase of the interference is simply given as

$$d_{SI} = (2p/l)BL, \quad B = |n_f - n_s|$$

where, B is the birefringent coefficient of the sensing fiber, L is the length of the sensing fiber, and n_f and n_s are the effective indices of the fast and slow modes, respectively [Fu, H.Y. et al 2008].

In general, high birefringent fibers (HBFs) or polarization maintaining fibers (PMFs) are chosen as the sensing fibers to acquire high phase sensitivity. For the temperature sensing application, the fiber is doped to have a large thermal expansion coefficient which induces high birefringence variation. When measuring other parameters such as strain, pressure, and twist, polarization-maintaining photonic crystal fibers (PMPCFs) are utilized as the sensing fibers. The merit of the SI-based sensors is the simultaneous sensing capability with the help of other fiber optic devices.

The interferometric fiber optic sensors have great potential in practical applications such as real time deformation monitoring of aircrafts, ships, bridges and constructions. Environmental sensors including explosive hydrogen sensor and biomedical sensors for health monitoring sensor systems are emerging fields. These novel sensors will evolve and expand their applications with the developments in specialty fibers and special fiber devices.

REVIEW QUESTIONS

1. What is electroptic effect? Explain its salient features.

2. Explain the importance of electroptic modulator and its applications.

3. Classify the various types of electroptic modulator.

4. Describe an electroptic device as a light beam modulator

5. Describe the acoustoptic device and its applications.

6. Describe acoustoptic device as a light beam modulator.

7. Describe acoustoptic device as a light beam deflector.

8. Write a note on fiber modulators.

9. What are the different types of interferometric sensors?

10. Explain in detail the various applications of interferometric sensors.

11. With necessary diagram, explain the Fabry–Perot interferometer liquid refractive index sensor.

12. Describe Mach–Zehnder interferometer with neat sketch and explain its working principle.

13. Describe a compact in-line Mach–Zehnder interferometer.

14. Describe the working of a sensor based on a Sagnac interferometer.

Chapter - IV

OPTICAL AMPLIFIERS AND FIBER OPTIC NETWORK COMPONENTS

4.1 OPTICAL AMPLIFIERS

To exploit the entire multi terahertz bandwidth of optical fibers, wavelength division multiplexing (WDM) sends signals of more than one wavelength through the same fiber. However, this important new technique would be impractical without optical fiber amplifiers - a technology now coming into widespread commercial use. By amplifying all signals in a broad window, without regard to wavelength or signal coding, fiber amplifiers allow expansion of carrying capacity of telecommunication networks without laying new fiber or installing new repeaters.

Any long distance optical communication network needs some sort of repeater to amplify signals enroute and to counteract the attenuation that inevitably occurs in even the most transparent fiber. In order to transmit signals over long distances (>100 km), it is necessary to compensate for attenuation losses within the fiber. Initially this was accomplished with an optoelectronic module consisting of an optical receiver, repeaters and detectors to send the data and is illustrated in Figure. 4.1. Such repeaters were limited by the electronic speeds at the time they were installed and had to be tailored to a particular transmission format and rate. Thus on terrestrial lines, they had to be reinstalled whenever capacity was to be increased this process is generally impractical with oceanic cable, so entirely new lines had to be laid in to increase capacity In addition, one repeater would be needed for each wavelength transmitted, making wavelength-division multiplexing an impossibly expensive proposition. That is, this arrangement is limited by the optical to electrical and electrical to optical conversions.

Optical amplifiers eliminate these problems because they amplify light wave signals without converting them to electronic pulses. Several types of optical amplifiers have been demonstrated to replace the optoelectronic – electronic regeneration systems. These systems eliminate the need for electrical to optical and optical to electrical conversions. This is one of the main reasons for the success of today's optical communications systems. The optical amplifiers are exceedingly simple in principle, and this chapter covers the recent advances in optical amplifiers in communications.

Is it possible to reduce the cost in these communications, is the commonly asked question. Semiconductor and other wave-guide based devices offer solution to this problem.

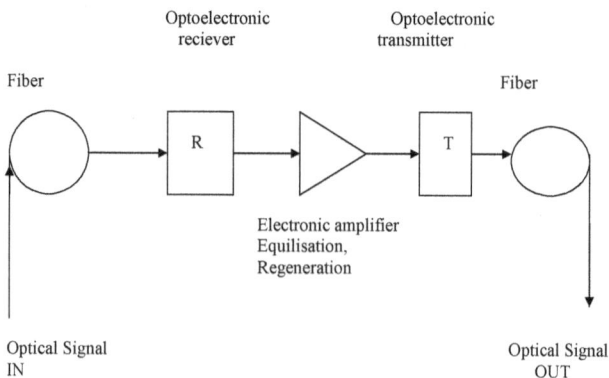

Figure 4.1 Schematic diagram of optoelectronic module consisting of an optical receiver and transmitter

A semiconductor optical amplifier (diode with antireflective facets) boosts signals by applying an electrical field to signal photons traveling along a wave guide structure that is usually based on InP, InGaAs or InGaAsP. Photons absorbed in the waveguide's material create an electron–hole pair that recombines to produce stimulated emission. If its emission rate outstrips its absorption rate, a semiconductor optical amplifier can provide gain as high as 15 dB (although it's typically lower).

4.2 BASICS OF OPTICAL AMPLIFIERS

Optical fiber amplifiers consist of an active medium that is pumped by an energy source into population inversion where higher energy states are more populated than lower ones. When the signal passes through the medium, the medium is stimulated to emit photons and hence signal is amplified. Although the signal makes only one pass through the medium, the amplification is by as much to 30–40 dB (a thousand fold to ten thousand fold). The general form of an optical amplifier is shown in Figure. 4.2.

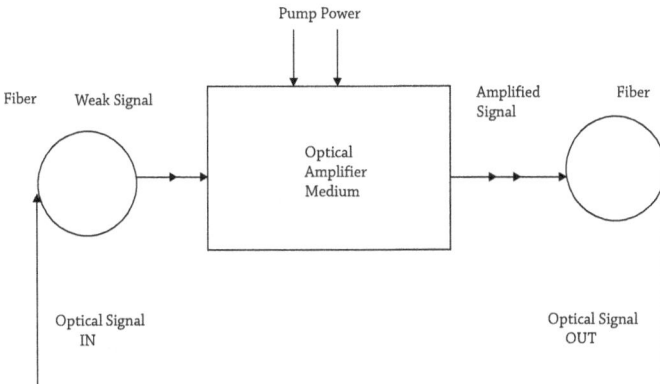

Figure 4.2   General form of an optical amplifier

Amplifiers occupy three locations in a long haul network each influencing design. The first one is the post or power amplifier located just after the transmitter end of a fiber span. Next comes, a series of line amplifiers located between each 80 km span of transmission fiber. Just before signals reach the receiver end, they are given a final boost by a preamplifier to improve incoming signal to noise ratio. The diagram shown in Figure. 4.3 illustrates pre, post and line amplifiers.

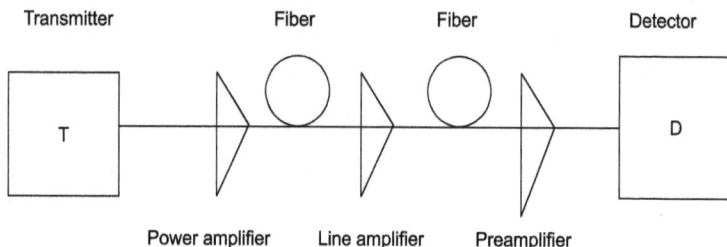

Figure 4.3   Pre, post and line amplifiers

There are three main types of optical amplifiers. They are

1. Optical fiber amplifiers (OFAs or active fibers).

2. Semiconductor optical amplifiers (SOAs). Semiconductor optical amplifiers are made up of semiconductor alloys involving phosphorous, gallium, indium and arsenic and are electrically pumped.

3. Rare earth doped fiber amplifiers (erbium – EDFA 1500 nm, praseodymium – PDFA 1300 nm). Fiber amplifiers use rare-earth elements – typically erbium, but also praseodymium, thulium, neodymium, and other – as dopants in a glass optical fiber and are optically pumped.

4. Fiber Raman and Brillouin amplifiers. Raman amplifiers use germanium doped fibers and high pumping powers to exploit nonlinearities, thus achieving gain. Both SOAs and Raman amplifiers have specialized use for telecommunications.

Fiber amplifiers have strong advantages – they are highly linear, are independent of polarization state and have low noise levels. The most practical optical amplifiers to date include the SOA and EDFA types. New pumping methods and materials are also improving the performance of Raman amplifiers.

4.2.1 Semiconductor Amplifiers

Semiconductor optical amplifiers are similar in configuration to the diode pump in a fiber or Raman amplifier that gives them an obvious edge in terms of cost. On the downside, these devices tend to have low gain due to (exacerbated) coupling losses and comparatively low channel capacity.

However, their chip-based designs allow integration of several functions with amplification, including gate switching, modulation and wavelength conversion (Photonic Integrated Circuit–PIC). These PICs make semiconductor devices likely candidates as integrated amplifiers that can serve as switch and amplifier in an add/drop multiplexer.

The drawback of semiconductor optical amplifiers is the peculiar nature of their response time that can make them sensitive to variances in signal power intrinsic to gigabit data rates. In effect, one data bit's gain can differ from the next, based on variations in the preceding bit's intensity and pulse length. The result is that many chip-based amplifiers can boost only one channel at a time without incurring cross talk.

This problem can be overcome by integrating a diode laser along the bottom of an InP-based wave-guide. Instead of pumping signal photons, the laser acts as ballast to allow them to traverse the chip's active layer without inducing variations in output gain. This provides the stability to handle multiple channels with a device one-hundredth the size of an erbium doped fiber amplifier.

Semiconductor Optical Amplifier (SOA) is similar to a laser cavity and is used as a discrete amplifier. They can be integrated into arrays of amplifying switching and gating devices. It finds application in all optical 3R–regeneration systems. The schematic diagram of a semiconductor optical amplifier is shown in Figure 4.4.

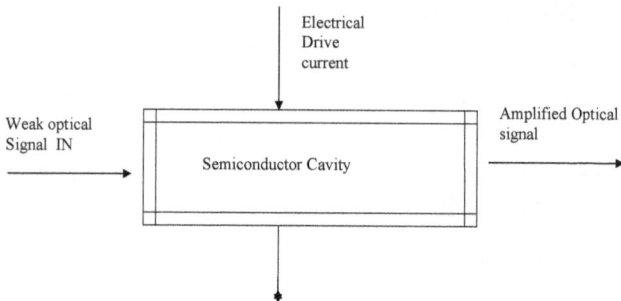

Figure 4.4   Schematic diagram of a semiconductor optical amplifier

Characteristics of SOA

1. They are polarization dependent and require polarization-maintaining fiber.

2. They possess relatively high gain (~20 dB).

3. Their output saturation power is 5–10 dBm.

4. They can operate at 800, 1300 and 1500 nm wavelength regions.

5. They are compact and easily integrated with other devices. They can be integrated into arrays.

6. They have high noise figure and cross-talk levels (due to nonlinear phenomenon such as 4–wave mixing). This feature restricts the use of SOAs.

7. SOAs are limited in operation below 10 Gb/s. (Higher rates are possible with lower gain.)

4.2.2 Rare earth doped fiber amplifiers

The key to practical application of optical amplifiers in telecommunications was in finding a dopant to match the wavelengths (1310 and 1550 nm) which are transparent to glass fibers (where all optical telecommunications takes place today). An attempt in this regard was successful in 1987. That is, an erbium-doped fiber amplifier (EDFA) amplified signals in a 3 THz wide band from 1530–1560 nm. One widely used type of optical amplifier is made from erbium-doped fiber Erbium-doped fiber amplifiers operates at 1550 nm and are pumped by diode lasers at 980 nm or 1480 nm. EDFA amplifier is used to boost the signal in the fiber optic communication systems. EDFA amplifier is a very important device that makes the WDM transmission possible, and it is the WDM that substantially increase the capacity of the fiber optic systems EDFA amplifier works on the principle of stimulated emission and the EDFA has two distinguishing features compared with the SOA. Its active medium is a piece of silica fiber heavily doped with ions of Erbium and its external energy is delivered in optical form. EDFA amplifier gain depends on the wavelength of the input signal. The main source of noise in an EDFA is the amplified spontaneous emission. An important feature of the EDFA amplifier in its use in WDM networks is its ability to keep the characteristics constant when one or more input channels are added or dropped. Rare earth doped fiber amplifiers are finding increasing importance in optical communications systems. Perhaps the most important version is erbium-doped fiber amplifiers (EDFAs) due to their ability to amplify signals at the low loss 1.55 μm wavelength ranges.

EDFAs have many advantages over semiconductor optical amplifiers (SOAs).

1. It has high power transfer efficiency from pump to signal power (> 50%).

2. It has wide spectral band amplification with relative flat gain (> 20 dB). Hence, it is useful for WDM applications.

3. Its high saturation output [greater than 1 mW (10 to 25 dBm)] and immunity to interference between different channels. This is due to the long fluorescence lifetime of the Er^{3+}–doped fiber.

4. Its gain-time constant is long (> 100 msec) to overcome patterning effects and inter-modulation distortions (low noise).

5.   It has large dynamic range.

6.   It has low noise figure, which is defined as the signal-to-noise ratio (SNR) degradation of the input signal.

7.   It is polarization independent.

8.   It is suitable for long-haul applications.

9.   Optical gain in EDFAs is intrinsically insensitive to signal polarization because the active Er^{3+}-ion dipoles are randomly oriented in the silica glass host matrix. This property is very important in lightwave communication systems because optical fibers do not preserve signal polarization.

10.   High-performance EDFAs have low nonlinear effects.

11.   Terrestrial networks, undersea-cable systems, WDM systems, and local lightwave networks using EDFAs have significantly increased transmission capacity, customer access, networking functionality and operational flexibility.

EDFAs also have some disadvantages

1.   They are relatively large devices (km lengths of fiber) and hence they cannot be integrated easily with other devices.

2.   They give rise to amplified spontaneous emission. There is always some output even with no signal input due to some excitation of ions in the fiber (spontaneous noise).

3.   They experience cross-talk effects and gain saturation effects. However, Erbium-doped fiber technology is nowadays very mature and is a cost–effective solution for conventional applications.

The energy level diagram for Er-doped silica is shown in Figure 4.5. Pumping is primarily done optically with the primary pump wavelengths at 1.48 µm and 0.98 µm. As shown in figure 4.5, atoms pumped to the 4I (11/2) level, 0.98 µm band decays to the primary emission transition band. Pumping with 1.48 µm light is direct to the upper transition levels of the emission band. Semiconductor lasers have been developed for both pump wavelengths. 10–20 mW of absorbed pump power at these wavelengths can produce 30–40 dB of amplifier gain. The pump efficiencies of 11 dB/mW are achieved at 980 nm. Pumping can also be performed at 820 and 670 nm with GaAlAs laser diodes but their pump efficiencies are lower. However, these lasers can be made with high output power.

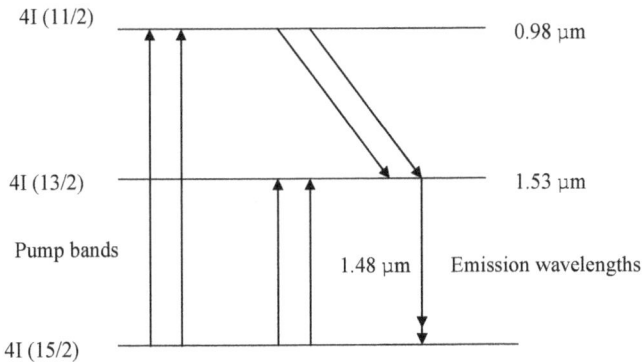

Figure 4.5 Energy level diagram for Er doped silica

Since the gain spectrum of erbium resembles a 3–level atom, it is possible to model the gain properties using this approach. Several different wavelength bands have been designated for wavelength division multiplexing. EDFAs are designed to operate in the following bands.

a. S–Band 1480–1520 nm,

b. C–Band 1521–1560 nm and

c. L–Band 1561–1620 nm

General EDFA Amplifier Configuration The general erbium doped fiber amplifier configuration is shown in Figure 4.6. The basic principle of erbium doped amplifier is, Signal Photons + Pump Photons + erbium atoms = more signal photons.

When pumped by 980 or 1480 nm laser, the energy level of erbium atoms jumps. Since atoms return to the ground state via spontaneous emission, it creates noise in the signal. However, most of them are knocked back to their ground state by signal photons with wavelengths somewhere between 1510 and 1600 nm. When this happens, the erbium ions emit a photon that has wavelength identical to signal, thus providing gain. The mechanics of this process works most effectively between 1535 and 1565 nm where erbium's gain curve forms a more or less flat plateau, a little better than –40 dBm. However, conventional erbium-doped amplifiers, today boost signals from 1530 to 1565 nm. Special type of fibers has even extended erbium-based amplifiers to operate in the L band from 1570 to 1620 nm.

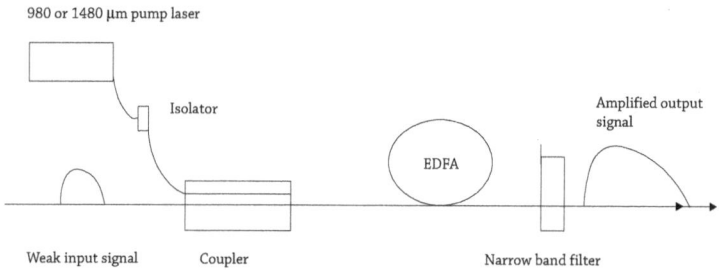

980 or 1480 μm pump laser

Isolator

EDFA

Amplified output signal

Weak input signal   Coupler   Narrow band filter

Figure 4.6   General erbium doped fiber amplifier configuration

Their basic characteristics are

1. Rare earth doped optical amplifiers work almost like a laser.

2. The primary difference between rare earth doped optical amplifier and laser is that the rare earth doped optical amplifier has no resonator.

3. Here, amplification occurs primarily through the stimulated emission process.

4. The medium is pumped until a population inversion state is achieved. Pump powers are typically several 20–250 mW. An isolator is used to reduce reflections at the input to the amplifier and a narrow band optical filter is used to reduce

transmission of amplified spontaneous emission frequency components.

5. The resultant optical gain depends both on the optical frequency and the local beam intensity within the amplifier section.

The gain coefficient of EDFA can be expressed as

$$g(\omega) = g_0 / \{1 + (\omega - \omega_0)^2 T_2^2 {}_+ P/P_s\}$$

where, g_0 is the peak gain,

ω is the optical frequency of the incident signal,

ω_o is the transition frequency,

P is the optical power of the incident signal,

T_2 is the dipole relaxation time and

P_s = is the saturation power.

Typically, T_2 is lesser than < 1 ps, and the saturation power P_s depends on gain medium parameters such as the fluorescence time and the transition cross section.

When not saturated (i.e. P/P_s <<1) the gain coefficient $g(\omega)$ becomes

$$g(\omega) = g_0 / \{1 + (\omega - \omega_0)^2 T_2^2\}$$

Gain is maximum when $\omega = \omega_0$ (i.e. the gain coefficient is at resonance). At non-resonant frequencies, the gain follows the homogeneously broadened characteristics of a two level atom (i.e. Lorentzian profile).

The gain bandwidth for this spectrum is typically expressed as the FWHM (Full Width at Half Maximum)

$$\Delta\omega_g = 2/T_2$$

$$\Delta\omega_g = \Delta\omega_g / 2\pi, \text{ with } T_2 \cong 0.1 \text{ ps}$$

$$\Delta v_g \cong 3 \text{ THz}$$

Large Spectral bandwidth amplifiers are preferred for fiber optic systems to make them less sensitive to dispersed transmitted signals and useful for WDM systems.

The gain spectrum of erbium ions alone is homogeneously broadened and the bandwidth is determined by the dipole relaxation time T_2. However when placed in a glass host the spectrum is influenced both by the silica and any other dopants. This can result in inhomogeneous broadening contributions.

The combined homogeneous and inhomogeneous bandwidth of EDFAs ~ 30 nm. The amplification factor of EDFA is defined as $G = P_{out}/P_{in}$, where P_{out} is the amplifier output power and P_{in} the input power of a CW input signal.

The amplification factor after a length L of optical amplifier medium is

$$G(\omega) = \exp[g(\omega)L]$$

Both $g(\omega)$ and $G(\omega)$ are a maximum when the frequency is at resonance $\omega = \omega_0$ and decrease when the frequency is detuned from resonance. However, the amplifier factor (G) decreases much faster than the gain coefficient (g).

Gain Saturation   Since $g(\omega)$ depends on the incident optical power when $P \approx P_s$, G will start to decrease with an increase in optical power P.

Assuming the incident frequency is tuned for peak gain $(\omega = \omega_0)$

$$dP/dZ = g_0 P/(1 + P/P_s)$$

With the conditions $P(0) = P_{inc}$ and $P(L) = P_{out} = GP_{inc}$ the large signal amplifier gain becomes

$$G = G_0 \exp\{-(G-1)P_{out}/G\,P_s\}$$

This expression shows how the amplifier gain decreases when $P_{out} \approx P_s$.

Output saturation power $\equiv$ the optical power at which G is reduced to $G_o/2$ (3 dB).

Amplifier Noise Spontaneous emission in the amplifier will degrade the signal to noise ratio (SNR) by adding to the noise during the amplification process.

SNR degradation is quantified through the amplifier noise figure F_n

$$F_n = (SNR)_{in}/(SNR)_{out}$$

where, the SNR is based on the electrical power after converting the optical signal to an electrical current. Therefore, F_n is referenced to the detection process and depends on parameters such as detector bandwidth (B_e) and thermal and shot noise.

Gain equalization in rare earth doped optical amplifier can be accomplished using

a. Thin film filters,

b. Long period fiber gratings and

c. Chirped fiber Bragg gratings.

Fiber amplifiers gain signal and not noise. Anyone who listens music from a stereo system is probably familiar with the hiss noise from the speakers between songs. The sound that comes from the amplifier electronics is usually forgotten with the first chords of the next track. That is because the signal power overcomes the amplifiers small noise factor. The similar thing happens in communication networks too. Here, they are much less tolerance of noise and require balanced design to maintain consistent signal quality. As a general rule, signals arriving at the receiver end should have between –10 and –5 dBm (0.1–0.3 mW) for each fiber channel and it should have a signal-to-noise ratio of 100:1. The next complexity increases due to receiver power, fiber non-linearity, network topology and a host of other factors. Since no two-network connections are alike, the amplifiers must be extremely adaptable in design. Since the mid 1990s, the erbium-doped amplifier (EDFA), helps in networks evolution with high power pump sources and gain flattening. The faster data rates, denser channel counts, expanded transmission bands or some combination of them demands improvements in pumps, equalizer components of erbium-based amplifiers along with new materials for fibers.

The most significant change that erbium doped amplifiers will encounter in long-haul networks is the increasing presence of Raman amplifiers. Erbium-doped fiber amplifiers have enabled dense wavelength division multiplexing (DWDM). Erbium's more or less flat gain between 1530 and 1565 nm simplifies the task of keeping decibel levels even from channel to channel. This peculiar characteristic of erbium was a serendipitous discovery and network designers quickly realised its potential. It could either extend the distance that a single channel could be transmitted without regeneration or support additional chance. Those who choose the latter option utilise the WDM revolution and establish erbium's flat stretch of gain as the conventional band, or C band.

Erbium in the L-band Higher power pumps serve erbium amplifiers well in C-band applications. Today's amplifiers typically support up to 80 channels. However, the demand for power from a single pump has plateaued as DWDM has begun expanding into the 1570–1620 nm range, the so-called long band or L-band, where erbium's gain drops. The reason is that it is easier to integrate several pumps into a module than to increase a single pump's output. Further, erbium has a low saturation threshold. We should also remember that after a point, erbium will not respond to more pump photons. Hence, a few argue that L–band amplification will rely on other technologies. As channel counts reach 100 and beyond, the amplifier's output must support so many dBs on each channel. Since EDFA's saturation level becomes a limitation, one has to depend on Raman amplifier whose efficiency is better. The level of pumping affects erbium's L-band gain profile. More power equals less bandwidth and vice versa. Therefore, we have to increase the gain medium (i.e., the erbium atoms), not the power. This is possible by lengthening the fiber component and/or by increasing the concentration of erbium dopant.

Basically, we need something about four to five times more (standard erbium-doped) fiber to get gain comparable to the C-band. This not only increases the amplifier's cost but also raises issues such as chromatic dispersion.

Let us also remember that mere increase of erbium atoms into a fiber is not going to solve the problem. At high concentrations, erbium ions cluster together and emit as many photons as would an individual ion. Or else the ions jump to higher levels and convert pump energy into signal.

One way around this is to dope the fiber with relatively high levels of aluminium to distribute erbium atoms more efficiently through the fiber matrix. Through this approach, Lucent produced L-band amplifier fiber components a little more than twice the length of those used in the C-band.

Until recently, erbium-doped fiber amplifiers were used for all long haul applications, serving as pre, post and line amplifiers. However, expanding network capacity and lengthening system including made the designers to search for alternative technologies. The Raman amplifiers solve the problem.

4.2.3 Raman Amplifiers

Before the discovery of erbium's gains properties, Raman amplifier was not considered in optical communication. The favourite semiconductor amplifier technology was very much appreciated as it avoided problems with cross talk, noise and polarization. Further, semiconductor diode lasers lacked sufficient power to stimulate the Raman effect in transmission fiber. However, early Raman devices amplified signals using double-clad ytterbium-doped fiber pumps which delivered around 3 W of output. Applications for these pumps continue even today in short underwater links where network providers want to avoid the expense of using undersea-qualified erbium-doped amplifiers.

Raman scattering is an inelastic scattering mechanism, which do not require a population inversion. Raman scattering is a weak effect. It occurs through a slight modulation of the refractive index through molecular vibrations of the material. The change in polarisability is an important condition for Raman effect. The scattered light with lower energy ($n_2 < n_1$) is stokes scattering while the scattered light with higher energy ($n_2 > n_1$) is anti-stokes Scattering. Stokes scattering typically dominates due to greater population of the ground state relative to the vibrational state when the system is in thermal equilibrium.

The molecules contributing to the process are vibrating independently and the scattered light is non-directional. This gives rise to Spontaneous Raman Scattering. At higher intensity levels, the generated photons begin to act in phase or coherently. i.e., the molecules oscillate as an array of vibrating oscillators. This gives rise to Stimulated Raman Scattering (SRS). SRS can be a problem but it can also be used as a signal amplification process. On the negative side, it contributes to dispersion and places an operational limit on the amount of power that can be transmitted through a fiber.

The Stokes wave is amplified as it propagates through the medium as

$$dI_2/dz = G_r I_2 I_1$$

Where, I_2 is the intensity of the stokes shifted light ($\omega_s = \omega_1 - \omega_{vib}$), I_1 is the intensity of the pump beam (ω_1) and G_r is the

Raman gain term that includes material factors. For $I_2 << I_1$ and cases where the pump beam is not significantly depleted

$$I_2(z) = I_2(0) \, e^{G_r \cdot I_1 \cdot z}$$

The properties of Raman amplifiers are

1. The peak resonance in silica fibers occurs about 13 THz from the pump wavelength. At 1550 nm this corresponds to a shift of about 100 nm.

2. The power is transferred from shorter wavelengths to longer wavelengths.

3. Coupling with the pump wavelength can be accomplished either in the forward or counter propagating direction.

4. Power is coupled from the pump only if the signal channel is sending a 1 bit.

An array of laser diodes can be used to provide the Raman pump. The beams are combined and then coupled to the transmission fiber. The pump beams can counter propagate to the direction of the signal beams. The pump arrangement to extend the range for stimulated Raman amplification is shown in Figure. 4.7.

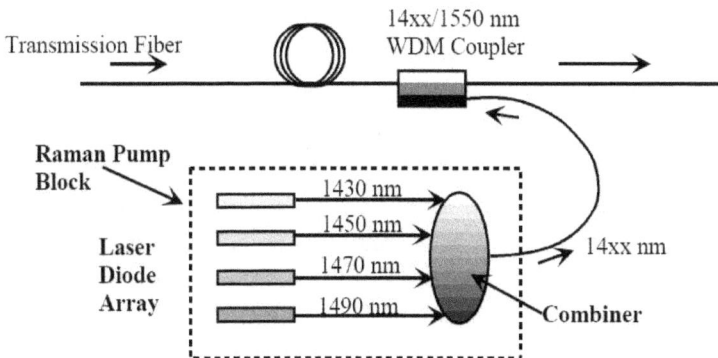

Figure 4.7 Pump arrangement for stimulated Raman amplification

Difficulties with Raman Amplifiers

1. The Pump and amplified signals are at different wavelengths. Therefore, the signal and the pump pulses will separate due to dispersion (waveguide dispersion) after a certain propagation distance. The difference in propagation time is given by

   $\delta t = (L/c)\lambda^2 d^2 n / d\lambda^2 (\delta v/v)$, where, L is the fiber length.

   A 1 psec pump pulse at 600 nm separates from a 1 psec stokes pulse in almost 30 cm.

2. The second problem is that the pump power decreases along the fiber length due to linear absorption and scattering while Raman gain is greater at the input end.

3. The third problem results from amplifying spontaneous Raman photons. This occurs when the pump power is increased to offset attenuation losses and spontaneous Raman photons are coupled into the guided mode all along the length of the fiber. This increases noise.

Upper limit on the power into a communications signal from SRS amplification can be defined as the point at which the Stokes power P_r equals the signal power P_{sig}.

$$P = 16\pi w_0^2/G_r L_{eff} \text{ where, } L_{eff} = (1-e^{-\alpha L})/\alpha$$

In the last two decades, diode laser pump powers have increased sufficiently (example – twin pumped design – two pairs of single mode 230 mW diode lasers) and provide more power to allow Raman amplifiers to play a vital role in fiber optics communication. More importantly, separating the pump wavelengths by 25 to 40 nm helps to keep the gain curve flat over a single band amplifier's 32-nm bandwidth.

Using Raman pump sources along with EDFAs, allows system architects to move to higher data rates and denser wavelength spacing more easily. In future, Raman technology is expected to become more prevalent and may even be used for 'stand-alone' amplification.

As for the future, it's likely that discrete Raman amplifiers will assume greater importance. In this case, a short length of specially tailored fiber boosts the signal – an arrangement similar to the EDFA.

At present, we have a Raman System that employs both distributed and discrete amplification to deliver a gain bandwidth of 132 nm around 1500 nm. There's plenty of work to be done before Raman amplification can become a mainstream technology. Some issues must be overcome if Raman amplification technologies are to be put to practical use.

For example,

1. Because Raman amplifiers need high pump powers, there is a safety problem if the transmission fiber breaks.

2. Pump lasers with sufficiently high powers are also not available for all wavelengths.

The amplification process adds noise causing unavoidable SNR degradation. The added noise impairs system performance by increasing the bit-error rate of the receiver in a lightwave system.

The spacing between pumps depends on the fiber we use, but typically we wish to use 1427 and 1467 nm pumps which provide gain in the C–band.

Another reason to use two pumps is that the Raman gain in a transmission fiber is polarization-dependence. The pump amplifies only signals that share the same polarization profile. Counter polarizing two pumps in a module allows the amplifier to overcome this drawback.

Raman's reach Besides fiber lasers, Raman amplifiers assume either Distributed Raman modules or Discrete Raman modules. The distributed Raman modules include pumps, optics and electronics and most often function as line amplifiers linking and amplifying fiber spans. It functions in conjunction with erbium-doped amplifiers. The discrete or lumped Raman modules are just "Raman in a box" – incorporate similar components as a distributed Raman amplifier plus a span of fiber designed for specific applications. This technology has merit in several ways.

1. Raman amplifiers improve the signal-to-noise number and allow faster bit rates.

2. It is a tool to open up transmission in a fiber's short band, or S-band between 1485 and 520 nm.

3. Raman amplifiers first burrowed into terrestrial applications and preamplifier applications.

Although EDFAs satisfied the requirements for communication (for a while), technologists wanted to upgrade the system due to developments in science and technology for better communications. Raman is a straightforward and cost-effective means to do that. The distributed Raman amplifier's main benefit is increasing system length (a system being defined as the distance between the electronic regenerators or repeaters that renew optical signal).

Raman amplification however is not going to solve all the problems. There are network design considerations such as the number of amplifiers and the type of fiber present. Smaller-core fiber requires lower pump power. Raman has a higher gain quotient in these fibers because their lower unit of area concentrates pumps power and enhances the Raman effect. In dispersion-compensation fibers, for example, Raman's gain can be 10 times as great.

If pumped too hard, the system encounters double signal Rayleigh backscatter, wherein some of the signal power bounces back in the direction of the pump and backscatters as second time (amplified and travels back and forth). The effect cascades in systems with distributed amplifiers, adding up from span to span. Consequently, systems with distributed Raman amplifiers tend to limit to approximately 15 dB per amp. The schematic diagram of Raman amplifier is given in Figure. 4.8.

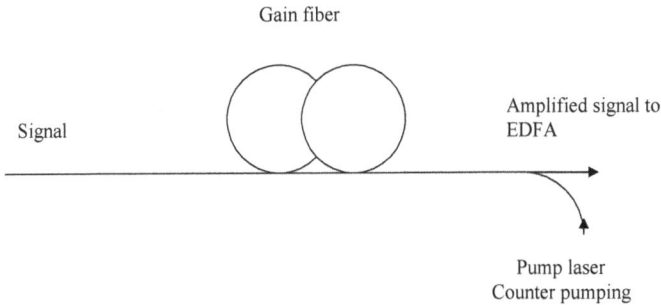

Gain fiber

Signal

Amplified signal to EDFA

Pump laser
Counter pumping

Figure 4.8 Schematic diagram of Raman amplifier

Except for silica's absorption peaks, Raman amplification is theoretically applicable over a single mode fiber's entire 300 to 2000 nm transmission band. It amplifies when pump photon couple with vibrational modes of a fiber's silica core. The transfer of energy is released in the form of photons shifted roughly 100 nm towards the red. In other words, Raman pump emitting at 1450 nm will provide gain at about 1550 nm. Raman amplifiers are counter propagating. That is, they amplify either the approaching or departing signal photons. Aimed against the signal, this boosts the approaching signals and directs the pump noise away from the receiver. This reducing its effect through fiber attenuation. Distributed Raman adds between 7 and 10 dB of signal strength per fiber span. This translates into 35 to 50 kms of extra reach, assuming that the loss of a typical transmission fiber is 0.2 dB/km.

Long on the short band The small-core fiber discrete or lumped Raman amplifiers are available in the market. Although discrete Raman amplifiers must pump, considerably harder to achieve equivalent gain, it opens applications in the S-band.

Let us remember that a substantial number of routes will need more than 160 channels. That means, we have to light more fibers which is expensive or go to 40 Gb which is not economical over 500 km. At that point, we will have the only option of expanding bandwidth and that's where the S-band comes in.

The S-band poses several potential panacea pitfalls for distributed Raman amplification.

1. No matter what band it amplifies, distributed Raman's powerful pumps can damage legacy fibers.

2. Another drawback for distributed amplification is that silica fibers have a water peak around 1380 nm that quickly attenuates pump energy.

3. Also, transmitting S-band signals leaves little room for distributed Raman pump sources to amplify the C- and L- bands.

However, some of these issues can be resolved by selecting wavelengths in the S-band that avoid the water peak in these legacy fibers and that locate pump wavelengths for L-band amplification separate from S-band signals. The discrete Raman amplifiers for the S-band do not rely on legacy fibers for the gain medium as does in distributed Raman technology. Hence, discrete Raman amplification in the S-band can be legacy fiber friendly.

Further, if we need an additional band onto a C-band network, it makes more sense to start with the S-band. That is because short–wavelength channels tend to lend some of their power to long-wavelength channels. Transmission signals in the S-band help to amplify the C-band and C-band signals help to boost the L-band. Therefore, progress on L-band onto a C–band network requires raising pump powers at the C-band amplifiers. Since the S-band will pump both the C- and L- bands (turning an amplifier down is easier than turning it up), it makes sense to add the L-band.

A typical profile for line amplifier in an 80 channel long haul system for EDFA and Raman amplifier is given in Table 4.1.

Table 4.1 EDFA versus Raman amplifier

	EDEA	**Raman amplifier (1450 nm)**
Gain	20 to 25 dB	10 dB
Bandwidth	32 nm	32 nm
Flatness	1 dB across bandwidth	1 dB across bandwidth
Output power	20 dBm or 100 mW	27 dBm or 500 mW
Power consumption	60 W	90 W
Noise figure	5.5 dB	−0.5 to −1 dB

4.2.4 Brillouin amplifiers

Raman and Brillouin scattering are fundamentally similar effects. Brillouin scattering is the interaction of light with acoustic waves in solids or liquids. The interaction occurs through the modulation of the refractive index of the medium that occurs in the alternating areas of compression and rarefaction of the acoustic wave. The acoustic wave thus forms a phase grating moving at the speed of sound in the medium. This grating can diffract an optical wave, changing its direction of propagation and its frequency via, the Doppler effect. In a single mode optical fiber, the only possible change of direction is back reflection and the frequency change is simply given by

$$\nu_B = \pm 2n(V_m/C)\nu_p$$

Where, n is the refractive index, V_m is the speed of sound in the medium, C is the speed of light and ν_p is the frequency of the optical wave or pump. The minus sign refers to the scattering from an acoustic wave propagating in the same direction as pump whereas the plus sign corresponds to the scattering from a wave counter propagating with pump.

Since signal amplification in fibers using stimulated Raman scattering requires hundreds of mW of pump power, it is natural to consider Brillouin amplification in which the same levels of gain can be achieved with two orders of magnitude less pump power. Stimulated Brillouin scattering is the non-linearity that occurs at the lowest optical power level in optical silica fibers. Hence, stimulated Brillouin scattering has seen as setting a limit on the launched power in lightwave systems. The narrower gain bandwidth can be turned to advantage by using stimulated Brillouin scattering as an optical filter as well as amplifier. However, the increased spontaneous emission of Brillouin scattering inevitably degrades the system performance and limits the applicability of stimulated Brillouin scattering for signal boosting in light wave systems.

4.3 COMMUNICATION COMPONENTS

4.3.1 Optical filters

Optical filters are devices that play important roles in photonics processing systems, optical communication and in many optical systems. In these filters, bandwidth and center wavelength can be tuned to desired regions and passband (lowpass, highpass, bandpass, and bandstop). These filters have potential applications in wavelength division multiplexing and dispersion equalization. Optical filters come under two categories viz., fixed and tunable optical filters. Tunable optical filters are essential components in a wide range of optical systems

Tunable optical filters with variable bandwidth and center frequency characteristics are important in applications where dynamic changes in the bandwidth and the center frequency of the filter are required. One such application is in frequency division multiplexing or multi-WDM optical systems which utilise the large bandwidth of optical fibers to increase the transmission capacity that is mainly limited by fiber dispersion. In WDM systems, tunable optical filters are used as optical demultiplexers at the receivers to select one or more desired channels at any wavelength. A few types of bandpass optical tunable filters are optical ring resonators, optical transverse filters and Fabry Perot interferometers etc. Another type of optical filter is the cascade coupler in which one port is used as a bandpass while the other port is used as a bandstop filter (Mach Zehnder channel adding or dropping filter).

They are also used to enable to tune wavelength in lasers and add/drop multiplexers which are used in optical data communication systems. The increasing demand for bandwidth in optical transmission systems has led to the further development of wavelength tunability in add/drop filters, light sources, semiconductor optical amplifiers and receivers. Consequently, tunability was analyzed for a broad variety of dispersive devices, including fiber Bragg gratings, MEMS-based Fabry–Perot filter, waveguides and diffraction gratings. A major application area for tunable optical filters is modern optical communication systems. They require continuously increasing transmission capacity which is nowadays provided by wavelength division multiplexing (WDM). Instead of installing new fiber links, additional individually modulated optical channels are added to already established transmission systems. These operate at fixed optical frequencies. With increasing channel numbers, high-quality tunable devices are becoming increasingly important for flexible network management. As a result, tunable filters represent key components in a large number of optical subsystems such as optical channel monitors, wavelength-selective add-drop multiplexers, and tunable lasers.

In addition, tunable optical filters enable the routing of optical signals which is currently performed in the electrical domain. Future fiber-optic networks will use this routing of individual wavelengths. This significantly reduces the need for costly and speed-limited electrical switching. The use of tunable filters involves a reduction in spare part holding and therefore may replace fixed filters. Recently, the use of tunable filters in optical communication modules has been transferred to analytical near-IR spectroscopy. Potential applications include detection of gases with absorption edges in the mid-IR and overtones in the near-IR spectral region. Conventional spectrometers are mostly based on scanning dispersive or grating/ diode array configurations which are sensitive to vibration and temperature. They also take up a large amount of space. Miniature spectrometers based on optical communication modules with integrated tunable optical filters do not have these disadvantages.

4.3.2   Attenuators

Attenuation is the reduction or loss of optical power as light travels through an optical fiber.

The longer the fiber is and the farther the light has to travel, the more the optical signal is attenuated. Single-mode fibers usually operate in the 1310 nm or 1550 nm regions, where attenuation is lowest. This makes single-mode fibers the best choice for long distance communications. Multimode fibers operate primarily at 850 nm and sometimes at 1300 nm. Multimode fibers are designed for short distance use. The higher attenuation at 850 nm is optical communication modules offset by the use of more affordable optical sources (the lower the wavelength, the less expensive the optics). The attenuation spectrum of optical fiber is shown in Figure. 4.9.

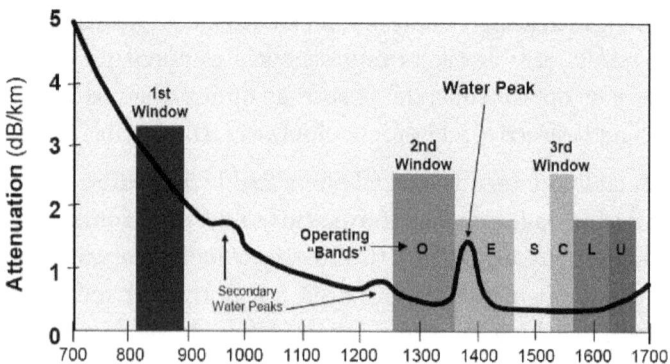

Figure 4.9   Attenuation spectrum of optical fiber (Courtesy: Furukawa Electric North America, Inc.)

A fiber optic attenuator (optical attenuator) simulates the loss that would be caused by a long length of fiber. While an optical attenuator can simulate the optical loss of a long length of fiber, it cannot accurately simulate the dispersion that would be caused by a long length of fiber. In a fiber optic receiver, too much light can overload it and degrade the bit error ratio. In order to achieve the best bit error ratio, the light power must be reduced. Fiber optic attenuators fit the requirement perfectly. This can happen when the transmitter delivers too much power such as when the transmitter is too close to the receiver.

Attenuators are similar to sunglasses which absorb the extra light energy and protect your eyes from being dazzled. Attenuators typically have a working wavelength range in which they absorb the light energy equally. An important characteristic of a good fiber attenuator is that they should not reflect the light, but they should absorb the extra light without being damaged. Since the light power used in fiber optic communications are fairly low, they usually absorb without noticeable damage to the attenuator itself. Fiber optic attenuators are available in 5dB, 10dB, 15dB and 20 dB attenuation values. For example, a –3 dB attenuator should reduce intensity of the output by 3 dB (50%). In attenuators, power rating refers to dissipated power.

Fiber optic attenuators are of two types namely fixed value attenuators and variable attenuators.

Fixed value attenuators have fixed values that are specified in decibels. The operating wavelength for optical attenuators should be specified for the rated attenuation because optical attenuation of a material varies with wavelength. Their applications include telecommunication networks, optical fiber test facility, and Local Area Network (LAN) and CATV systems. Further, fixed value attenuators are composed of two groups. They are 1) In-line type and 2) connector type. In-line type looks like a plain fiber patch cable. The connector type attenuator looks like a bulk head fiber connector. It has a male end and a female end that mates to regular connectors of the same type.

Variable attenuators come with many different designs. In general, they are used for testing and measurement. However, they are also used in EDFAs for equalizing the light power among different channels.

4.3.3 Circulators and isolators

Circulators and isolators are key elements in modern VHF, UHF and microwave engineering. Their fundamental property of non-reciprocity is capable of simplifying the construction and improving the stability, efficiency and accuracy of radar, communication and testing systems and industrial heating applications. The devices contain a core of ferrite material biased by a static magnetic field. This field orients the electron spins within the ferrite to produce a gyromagnetic effect. The non-reciprocal behaviour occurs when a RF signal, applied perpendicular to the biasing field, interacts with the precessing electrons to set up a standing-wave pattern within the core.

Circulators A circulator is a passive non-reciprocal device with three or more ports. Energy introduced into one port is transferred to an adjacent port, the other ports being isolated. Although circulators can be made with any number of ports, the most commonly used are 3-port and 4-ports.The symbols for 3-port circulator are given in Figure 4.10.

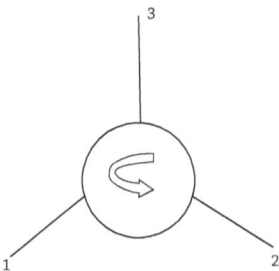

Figure 4.10 Symbol for 3 port circulator

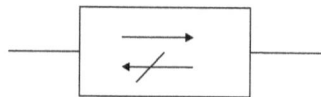

Figure 4.11 Symbol for an isolater

Energy entering into port 1 emerges from port 2; energy entering into port 2 emerges from port 3, and so on in cyclic order. In a circulator, the maximum power is the largest power it can handle at sea level and at maximum ambient temperature when one port is terminated with a mismatch giving a voltage standing wave ratio (VSWR) of 2 whilst the next port is matched with a VSWR of 1.2 or less, unless otherwise stated. This power value must not be exceeded. If the mismatch of the load is expected to exceed a VSWR of 2, a circulator of higher power handling capacity should be used. The maximum power is the maximum continuous-wave power unless a maximum peak power is separately stated. If this value is exceeded, the circulator can be damaged by arcing in its internal transmission structure. Power values are valid for one signal passage only. If more than one signal passes through the circulator, the peak power of the combined signal should not exceed the indicated maximum peak power. A common application for a circulator is as an inexpensive duplexer (a transmitter and receiver sharing one antenna).

Isolators   Isolators are common place in laboratory applications to separate a device under test (DUT) from sensitive signal sources. An isolator is a passive non-reciprocal 2-port device which permits RF energy to pass through it in one direction whilst absorbing energy in the reverse direction. An isolator is a circulator with the third port terminated The symbol for an isolator is given in Figure. 4.11.

In a circulator, isolation is the ratio is the input power to the output power for signal injection in the reverse direction and expressed in dB. Isolators are essentially the same design as a circulator except that one of the ports is terminated resulting in only one output. Isolators are installed so that they "Isolate" the output. This isolation prevents damaging standing waves from getting back to the transmitter's output stage, which would result in a transmitter circuit's failure.

In an isolator, the maximum power is the largest power that may be passed through it in the forward direction into a load with a VSWR of 2, unless otherwise stated. This power value must not be exceeded. Fig. 4.12 illustrates the most common application for an isolator. The isolator is placed in the measurement path of a test bench between a signal source and the device under test (DUT) so that any reflections caused by any mismatches will end up at the termination of the isolator and not back into the signal source.

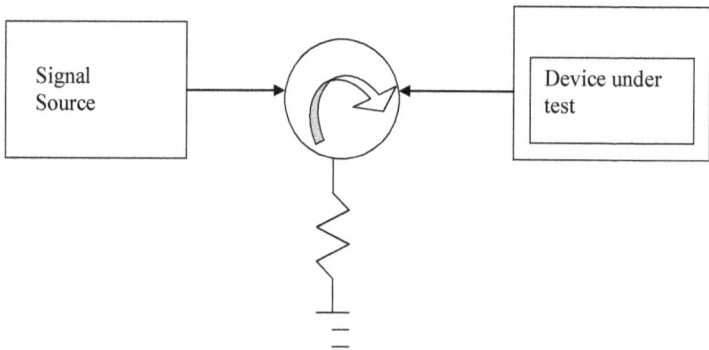

Figure 4.12   Common application for an isolator

When greater isolation is required, a dual junction isolator is used. A dual junction isolator is effectively two isolators in series but contained in a single package. Typical isolation performance can range from 40 to 50 dB with this type of design.

An important consideration when specifying an isolator or circulator is to ensure the device has adequate isolation for the given application. The greater the isolation value, the less interference from a signal on one port is present at the other. The amount of isolation is directly affected by the VSWR presented at port 3 of the isolator. If the match on port 3 is poor, one can expect isolation below 10 dB, but if the match is improved to 1.10:1 by using a good termination device in the circuit then the isolation would improve to over 20 dB.

Another important consideration when specifying circulators and isolators is to ensure the device has minimal insertion loss when inserted in a transmission path. Generally, the insertion loss of a circulator/isolator becomes more significant at higher frequency, namely because loss increases with frequency and higher frequency power sources are considerably more expensive. Accordingly, the criteria of low insertion loss will prevent precious power from being wasted.

4.3.4 OPTICAL SWITCHES

Fiber-optic switching device maintains the signal as light from input to output. Electro-optic was the earlier optical switch that connected optical fiber lines. They convert photons from the input side to electrons internally in order to do the switching and then convert back to photons on the output side. Truly optical switches support all transmission speeds. Optical switches direct the incoming bit stream to the output port no matter what the line speed or protocol (IP, ATM, SONET) and do not have to be upgraded for any such changes. Optical switches may separate signals at different wavelengths and direct them to different ports. In fiber optics, an optical switch is a switch that allows for signals that are traveling along one line of fiber to be switched selectively from one single circuit to another.

Using tiny mirrors that reflect the input signal to the output port, MEMS technology is expected to be the prevailing method for building optical switches, also known as "photonic switches". An optical switch is the unit that actually switches light between fibers and a photonic switch is one that does this by exploiting nonlinear material properties to steer light (i.e., to switch wavelengths or signals within a given fiber).

There are many different types of optical switches. They all serve the exact same purpose namely to switch data transfer from one fiber to another but they differ on how the change is made and how fast the change is made. Here are the most common types of optical switches:

a. Mechanical This means that each individual fiber is manually moved and/or moved together to ensure that the switch takes place.

b. Electro-optic Effects This is a change in the electric field that varies slowly in comparison to the speed of light.

c. Magneto-optic Effects By the use of electromagnetic forces, the switch can be made to transfer the data from one fiber to the other.

Semiconductor optical switches The unique properties of light is extremely helpful in transmitting several channels very fast and at the same time in fiber optic cables. Semiconductors play a vital role in turning the light beam on and off (digital 1s and 0s) in line with information available in the light beam. Unfortunately the speed in which the transformation takes in semiconductor switches is slower than light. This is one of the limitation faced in semiconductor optical switches in the early days. This difficulty was overcome by special crystals which operate at the speed of light. The current technology advanced in such a way to change this modern semiconductor switch on a chip. Hence, the use of semiconductor materials to switch light is not at all a constraint in fiber optics communication. Now, light beam with semiconductor optical switches function with speed of 0.3 picoseconds which is close to the speed of light. This is another milestone in fiber optics communication.

4.3.5 COUPLERS AND SPLITTERS

Fiber couplers or splitters are special fiber optic devices with one or more input fibers for distributing optical signals into two or more output fibers. The optical light is passively split into multiple output signals (fibers), each containing light with properties identical to the original except for reduced amplitude. Because the splitter is a passive device it is immune to EMI, consumes no electrical power and does not add noise to system design. The splitter's passive design is bi-directional and operationally independent of wavelength...

Fiber couplers have input and output configurations defined as M × N. M is the number of input ports and is one or greater and N is the number of output ports and is always equal to or greater than M. When there are multiple inputs, output signals are always a combination of the input signals – a coupler can also be considered a combiner.

Fiber optic couplers or splitters are available in a wide range of styles and sizes to split or combine light with minimal loss. All couplers are manufactured using a very simple proprietary process that produces reliable, low-cost devices. They are physically rugged and insensitive to operating temperatures.

Fiber, connectors, and splices rank as the most important passive devices. However, closely following are tap ports, switches, wavelength-division multiplexers, bandwidth couplers and splitters. These devices divide, route, or combine multiple optical signals. Some of the most common applications for couplers and splitters include

1. Local monitoring of a light source output (usually for control purposes).

2. Distributing a common signal to several locations simultaneously. An 8-port coupler allows a single transmitter to drive eight receivers.

3. Making a linear, tapped fiber optic bus. Here, each splitter would be a 95%–5% device that allows a small portion of the energy to be tapped while the bulk of the energy continues down the main trunk.

Couplers   Fiber optic couplers either split optical signals into multiple paths or combine multiple signals on one path. Optical signals are more complex than electrical signals, making optical couplers trickier to design than their electrical counterparts. Like electrical currents, a flow of signal carriers, in this case photons, comprise the optical signal. However, an optical signal does not flow through the receiver to the ground. Rather, at the receiver, a detector absorbs the signal flow. Multiple receivers, connected in a series, would receive no signal past the first receiver which would absorb the entire signal. Thus, multiple parallel optical output ports must divide the signal between the ports, reducing its magnitude. The number of input and output ports, expressed as an M × N configuration, characterizes a coupler. The letter M represents the number of input fibers, and N represents the number of output fibers. Fused couplers can be made in any configuration, but they commonly use multiples of two (2 × 2, 4 × 4, 8 × 8, etc.).

Splitters   The simplest couplers are fiber optic splitters. These devices possess at least three ports but may have more than 32 for more complex devices. Figure. 4.13 illustrates a simple 3-port device, also called a tee coupler. It can be thought of as a directional coupler. One fiber is called the common fiber, while the other two fibers may be called input or output ports. The coupler manufacturer determines the ratio of the distribution of light between the two output legs. Popular splitting ratios include 50% to 50%, 90% to 10%, 95% to 5% and 99% to 1%. These values are sometimes specified in dB values. For example, using a 90% to 10% splitter with a 50 μW light source, the outputs would equal 45 μW and 5 μW. However, excess loss hinders that performance. All couplers and splitters share this parameter. Excess loss assures that the total output is never as high as the input. Loss figures range from 0.05 dB to 2 dB for different coupler types. An interesting, and unexpected, property of splitters is that they are symmetrical. For instance, if the same coupler injected 50 μW into the 10% output leg, only 5 μW would reach the common port.

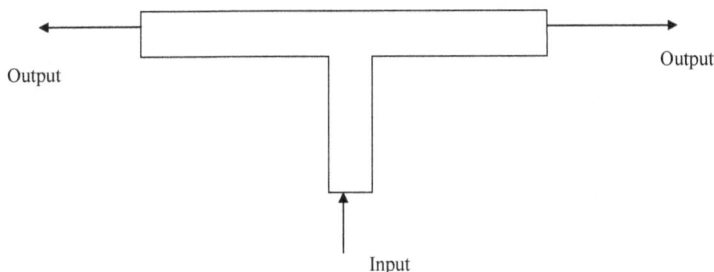

Figure 4.13 Typical Tee Coupler

Coupler and Splitter Applications In applications that require links other than point-to-point links, optical couplers find the widest use. This includes bidirectional links and local area network (LAN). In LAN applications, either a star network topology or a bus topology incorporates couplers. One use of a star coupler creates a large party-line circuit. Many transceivers connect to the star coupler and can communicate with all other transceivers, assuming the network adopts a protocol which prevents two or more transceivers from communicating simultaneously. Large insertion loss, (20 dB typically for a 64-port device) creates the biggest disadvantage of the star coupler. A bus topology may operate in a single direction or a bidirectional or duplex transmission configuration. In a one way, unidirectional setup, a transmitter at one end of the bus communicates with a receiver at the other end. Each terminal also contains a receiver. Duplex networks add a second fiber bus or use an additional directional coupler at each end and at each terminal. In this way, signals flow in both directions. The most popular coupler is a fused fiber coupler. In this coupler, two or more fibers are twisted together and melted in a flame.

Fiber optic couplers are widely used in feedback control circuits, ethernet and automotive LANs, medical instruments, automotive electronics, optical sensors, wavelength multiplexing, audio systems and communications systems

4.3.6 WAVELENGTH CONVERTERS

Wavelength conversion is used in WDM networks to improve efficiency. To understand wavelength conversion, consider the example shown in Figure. 4.14.

Two light paths proposed in the network are as follows.

1. Between node 1 and node 2 on wavelength λ_1 and

2. between node 2 and node 3 on wavelength λ_2

Let us assume that a light path has to be set up between Node 1 and Node 3. Such an act is impossible even though there is a free wavelength on each of the links along the path from node 1 to node 3. This is because the available wavelengths on the two links are different. Hence, a wavelength continuity network may suffer from higher blocking as compared to a circuit-switched network.

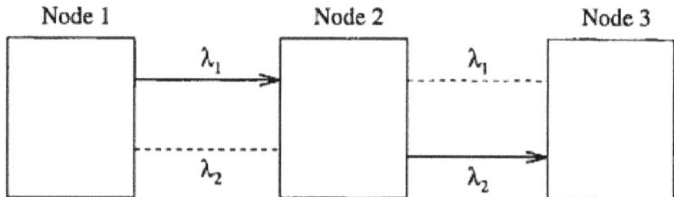

(a) without converter

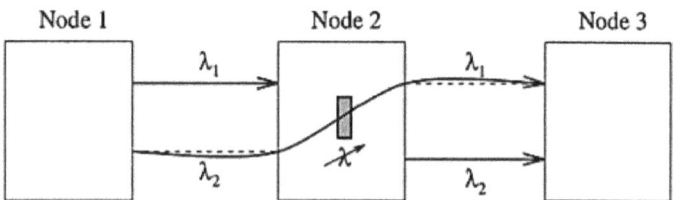

(b) with converter

Figure 4.14 Wavelength-Continuity Constraint in a Wavelength-Routed Network

It is easy to eliminate the wavelength-continuity constraint, if we were able to convert the data arriving on one wavelength along a link into another wavelength at an intermediate node and forward it along the next link. Such a technique is referred to as *wavelength conversion*.

In Figure 4.14, a wavelength converter at node 2 is employed to convert data from wavelength λ_2 to λ_1. The new lightpath between node 1 and node 3 can now be established by using the wavelength λ_2 on the link from node 1 to node 2 and then by using the wavelength λ_1 to reach node 3 from node 2. A single lightpath in such a wavelength-convertible network can use a different wavelength along each of the links in its path. Thus, wavelength conversion may improve the efficiency in the network by resolving the wavelength conflicts of the lightpaths.

Wavelength Converter Functions The function of a wavelength converter is to convert data on an input wavelength onto a possibly different output wavelength among the N wavelengths in the system. The functionality of a wavelength converter is shown in Figure. 4.15.

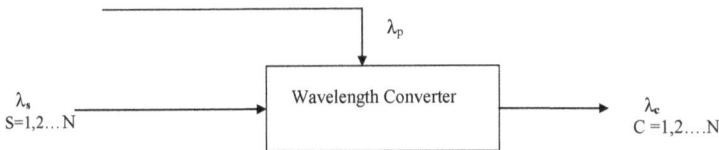

Figure 4.15 Functionality of a Wavelength Converter

Here, λ_s is the input signal wavelength, λ_c is the converted wavelength, λ_p is the pump wavelength, f_s is the input frequency, f_c is the converted frequency, f_p is the pump frequency and CW is the continuous wave (unmodulated) generated as the pump signal.

An ideal wavelength converter should possess the following characteristics

1. Transparency to bit rates and signal formats

2. Fast setup time of output wavelength

3. Conversion to both shorter and longer wavelengths

4. Moderate input power levels

5. Possibility for same input and output wavelengths (i.e., no conversion)

6. Insensitivity to input signal polarization

7. Low-chirp output signal with high extinction ratio and large signal-to-noise ratio

8. Simple implementation

Four fundamental approaches are adopted for wavelength conversion. They are

a. Optoelectronic,

b. Optical Gating,

c. Interferometric and

d. Wave Mixing

The latter three approaches are all-optical but not yet mature enough for commercial use. Optoelectronic converters today offer substantially better performance at lower cost than comparable to all-optical wavelength converters. For all-optical wavelength converters, polarization–dependent loss should also be kept to a minimum.

Optoelectronic Approach In optoelectronic wavelength conversion, the optical signal to be converted is first translated into the electronic domain using a photo detector. The electronic bit stream is stored in the buffer (FIFO for First-In–First-Out queue mechanism). The electronic signal is then used to drive the input of a tunable laser (T) tuned to the desired wavelength of the output. A typical optoelectronic wavelength converter is shown in Figure 4.16.

Address Decoder

λ_s → R → FIFO → T → λ_c

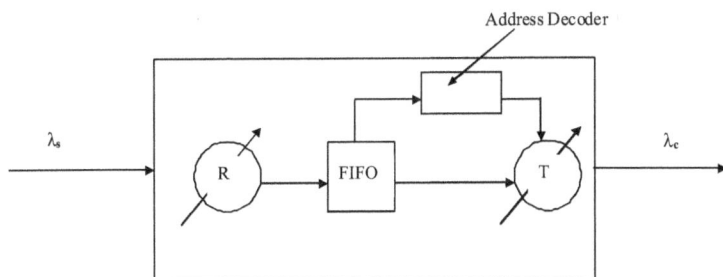

Figure 4.16 A typical Opto–Electronic Wavelength Converter

This is perhaps the simplest, most obvious and most practical method used for wavelength conversion. This is usually a variable-input with fixed-output converter. The receiver does not care about the input wavelength, as long as it is in the 1310 or 1550 nm window (fixed–wavelength laser). A variable output can be obtained by using a tunable laser. The performance and transparency of the converter depend on the type of regeneration used.

Optoelectronic converters offer only limited digital transparency. Moreover, deploying multiple optoelectronic converters in a WDM intermediate node requires sophisticated packaging to avoid crosstalk among channels. This leads to increased cost per converter and hence this technology is less attractive than all-optical converters.

Other disadvantages of optoelectronic converters include complexity and large power consumption. There are a lot of various schemes developed for all optical wavelength conversion in which each has its own strength and weakness in application. The recent trend in all-optical converters is the usage of periodically-poled LiNbO$_3$ waveguide for wavelength conversion scheme.

4.4 FIBER OPTIC TRANSMITTERS AND RECEIVER

The fiber optic transmitter is a device which includes a LED or a laser source and signal conditioning electronics that is used to inject a signal into fiber. The Fiber optic receivers capture the light from a fiber optic cable, decode the binary data it is sending and then convert into an electrical signal. Information is sent from a source to a transmitter by means of an electrical signal. The transmitter then takes that binary data and transfers it to a light signal. The light signal is passed through fiber optic cables and connectors until it reaches the receiver. The receiver then takes that light signal, translates it back to an electrical signal allowing the binary data to be read by the user. A transceiver shown in Figure. 4.17 is a device which combines the functions of both the transmitter and receiver.

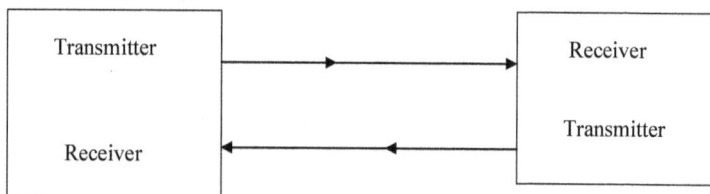

Figure 4.17   Fiber Optic transceiver (Data link)

In Fiber optic transmission systems, a transmitter is on one end of a fiber and a receiver on the other end. Most systems operate by transmitting in one direction on one fiber and in the reverse direction on another fiber for full duplex operation. In transceiver, both transmission and receiver are in a single module. The transmitter takes an electrical input and converts it to an optical output from a laser diode or LED. The light from the transmitter is coupled into the fiber with a connector and is transmitted through the fiber optic cable plant. The light from the end of the fiber is coupled to a receiver where a detector converts the light into an electrical signal which is then conditioned properly for use by the receiving equipment.

4.4.1   Sources for Fiber Optic Transmitters

The sources used for fiber optic transmitters should have correct wavelength, able to be modulate fast enough to transmit data and couple efficiently into fiber. The commonly used sources are LEDs, Fabry–Perot (FP) lasers, distributed feedback (DFB) lasers and vertical cavity surface-emitting lasers (VCSELs). LEDs and VCSELs are fabricated on semiconductor wafers such that they emit light from the surface of the chip while f–p lasers emit from the side of the chip from a laser cavity created in the middle of the chip.

As we know, LEDs have much lower power outputs than lasers. Their high diverging light output pattern makes them harder to couple into fibers, and hence it is used with multimode fibers. LEDs have much less bandwidth than lasers and are limited to systems operating up to about 250 MHz or around 200 Mb/s. LEDs have a very broad spectral output which causes them to suffer chromatic dispersion in fiber. On the other hand, lasers are easily coupled to single mode fibers and hence used for long distance high-speed links. Lasers have very high bandwidth capability and useful over 10 GHz or 10 Gb/s. Lasers have a narrow spectral output that suffers very little chromatic dispersion. LEDs and VCSELs are cheap to fabricate while lasers are more expensive due to the fabrication of the laser cavity inside the device which is more difficult. Every time the chip must be separated from the semiconductor wafer and each end should be coated even to test the chip for good laser. Typical fiber optic source specifications are given in Table 4.2.

Table 4.2 Typical Fiber Optic Source Specifications

Device	Wavelength (nm)	Power in to fiber (dBm)	Bandwidth	Fiber type
LED	850, 1300	−30 to −10	<250 MHz	MM
Fabry–Perot Laser	850, 1310 (1280–1330),1550 (1480–1650)	0 to +10	>10 GHz	MM, SM
DFB Laser	1550(1480–1650)	0 to +25	>10 GHz	SM
VCSEL	850	−10 to 0	>10 GHz	MM

DFB lasers are used in long distance and DWDM systems as they have the narrowest spectral width which minimizes chromatic dispersion on the longest links. Further, the light outputs of DFB lasers are highly linear as it directly follows from the electrical input. Hence, they are used as sources in amplitude modulation Community Antenna Television (AM CATV) systems. The choice of these devices is determined mainly by speed and fiber compatibility issues. Here, the electronics convert an incoming pulse (voltage) into a precise current pulse to drive the source. Lasers generally are biased with a low DC current and modulated above that bias current to maximize speed.

4.4.2 Detectors for Fiber Optic Receivers

Receivers use semiconductor detectors such as photodiodes, or photo detectors [positive–intrinsic–negative (PIN) or avalanche photodiode (APD)] to convert optical signals to electrical signals. Silicon photodiodes are used for short wavelength links, while InGaAs is used for long wavelength links. Very high speed systems sometimes use avalanche photodiodes (APDs) that are biased at high voltage to create gain in the photodiode. These devices are more expensive and more complicated to use but offer significant gains in performance.

4.4.3 Link Design and Optical link Power Budget

Link design consists basically of two functions

1. calculating optical power losses occurring between the light source and the photo detector and

2. determining bandwidth limitations on data carrying abilities imposed by the transmitter, fiber and receiver.

Reductions in optical power loss, or attenuation, as the light pulse travels through the fiber are expressed in dB/km (decibels per kilometer). A 10 dB loss means that 1/10 of the power arrives at the receiver, a 90% loss. Fiber optic links can operate with as little as 1/1000 of the output powers being received at the other end (a 30 dB loss). If the source emits sufficient power and the receiver is sensitive enough, the system can operate with high losses. How much loss can be tolerated will be determined by the stated minimum requirements of the receiver selected.

The prime causes of optical attenuation in fiber systems are

a. coupling loss,

b. optical fiber loss,

c. connector loss and

d. splice loss

The sum of the losses of each individual component between transmitter and receiver comprise the Optical Link Power Budget shown in Table 4.3.

Table 4.3   Typical Optical Link Power Budget

	Actual Power	**Optical Power Level**
Minimum optical power required by the receiver	0.1 μW	−40 dBm
Source output optical power	1 mW	0 dBm
Total operating budget (optical power)		40
SNR voltage ratio required in the receiver is 36 dB.		18 dB
The equivalent optical power ratio is dB = 20 log V1 / V2		
Remaining optical power for link		22 dB
Link optical power losses:	Excess Budget	2 dB
Cable 15 dB		
Connectors 3 dB		
Couplings 2 dB		
Total 20 dB		

The designer must consider these losses and select a transmitter and receiver combination that will deliver enough power to faithfully reproduce the signal. Some safety margins should also be made to future repairs or splices to the system, and age degradation of the source emitter. For example, a 3 to 6 dB margin for repairs and aging of the emitter is commonly employed.

4.5 WAVELENGTH-DIVISION MULTIPLEXING(WDM) AND DENSE WAVELENGTH-DIVISION MULTIPLEXING (DWDM)

The term wavelength-division multiplexing is commonly applied to an optical carrier described by its wavelength. Wave Division Multiplexing (WDM) is a technology that paves a way for the fiber optic cable to double its available bandwidth. WDM is a technology based on photonics. Photonics provide a method to send bits of information using pulses of light through fiber optic cables. WDM can generate many channels through a single fiber, allowing each channel to carry different bits of network traffic. The increase in bandwidth on existing fiber optic cable helps the users to achieve higher bandwidths without laying additional cable.

In fiber-optic communications, WDM is a technology which multiplexes a number of optical carrier signals onto a single optical fiber by using different wavelengths of laser light. This technique enables bidirectional communications over one strand of fiber as well as increasing capacity.

On the other hand, dense wavelength division multiplexing (DWDM) is a technology that puts data from different sources together on an optical fiber. Using DWDM, up to 80 channels (Wavelengths) of data can be multiplexed into a light stream transmitted on a single optical fiber.

4.5.1   Why WDM?

WDM is necessary to

1. Upgrade the capacity of existing fiber networks without adding fibers,

2. Transparency– Each optical channel can carry any transmission format (different asynchronous bit rates, analog or digital)

3. Wavelength routing (Wavelength is used as another dimension to time and space) and switching, and

4. For additional demand, one can buy and install equipment easily.

Each wavelength is like a separate channel and passive/active devices are used to combine, distribute, isolate and amplify optical power at different wavelengths. WDM technology is implemented in three variations as narrow band, wide band and dense band. Narrowband WDM (NWDW) is a technology that doubles the fiber span's capacity. NWDM implements two wavelengths, typically 1533 and 1577 nm wavelengths. Wideband WDM (WWDM) is technology that increases a fiber span by twofold. WWDM is implemented by combining a 1310 nm wavelength with another wavelength into the low-loss window of a fiber optic cable between 1528 nm and 1560 nm in wavelength.

This technology can further be divided into wavelength pattern as Conventional or Coarse WDM (CWDM) and Dense WDM (DWDM). WDM, DWDM and CWDM are based on the same concept of using multiple wavelengths of light on a single fiber but they differ in the spacing of the wavelengths, number of channels and the ability to amplify the multiplexed signals in the optical space.

WDM technology uses multiple wavelengths to transmit information over a single fiber. First WDM networks used just two wavelengths – 1310 nm and 1550 nm. CWDM has wider channel spacing (20 nm) at low cost. CWDM is also being used in cable television networks where different wavelengths are used for the downstream and upstream signals. In these systems, the wavelengths used are often widely separated. Passive CWDM is an implementation of CWDM that uses no electrical power. It separates the wavelengths using passive optical components such as bandpass filters and prisms.

Dense WDM (DWDM) is a technology that can increase a fiber span's capacity by eightfold. The DWDM has dense channel spacing (0.8 nm) which allows simultaneous transmission of 16+ wavelengths with high capacity. This fall into two different bands, a red band and a blue band. Each band is used in opposing directions. Today's DWDM systems utilize 16, 32, 64, 128 or more wavelengths in the 1550 nm window. Each of these wavelengths provides an independent channel. The range of standardized channel grids includes 50, 100, 200 and 1000 GHz spacing. In DWDM, discrete wavelengths form individual channels which can be modulated, routed and switched individually. This requires variety of passive and active devices. Wavelength spacing practically depends on laser linewidth and optical filter bandwidth.

A WDM system uses a multiplexer at the transmitter to join the signals together and a demultiplexer at the receiver to split them apart. With the right type of fiber, it is possible to have a device that does both simultaneously and can function as an optical add-drop multiplexer. The first WDM systems combined only two signals. Modern systems can handle up to 160 signals and can thus expand a basic 10 Gbit/s system over a single fiber pair to over 1.6 Tbit/s.

WDM devices have two components namely passive optical components and active optical components. The passive optical components consist of wavelength selective splitters and wavelength selective couplers. These operate completely in the optical domain (no O/E conversion) and do not need electrical power. They use fiber based or optical waveguides based or micro (nano) optics based technology. They can be fabricated using optical fiber or waveguide (with special material like InP, LiNbO3). They split or combine light stream using N × N couplers, power splitters, power taps and star couplers.

The active optical components comprises of tunable optical filter, tunable source, optical amplifier, add-drop multiplexer and de-multiplexer. Nortel's WDM System is shown in Figure. 4. 18.

Figure 4.18   Nortel's WDM System (Courtesy: Wikipedia)

Constraints in WDM Networks

1. The non-linear inelastic scattering processes due to interactions between light and molecular or acoustic vibrations in the fiber causes Stimulated Raman Scattering (SRS) and Stimulated Brillouin Scattering (SBS).

2. The non-linear variations in the refractive index due to varying light intensity causes self phase modulation (SPM), cross phase modulation (XPM) and four wave mixing (FWM).

4.5.2 DWDM

DWDM not only increase bandwidth but also combines and transmits multiple signals simultaneously at different wavelengths on the same fiber. That is, the single fiber is transformed into multiple virtual fibers. DWDM is independent of protocol and bit rate. DWDM based networks can transmit data in IP, ATM, SONET/

SDH and Ethernet. It handles bit rates between 100 Mb/s and 2.5 Gb/s. Therefore, DWDM-based networks can carry different types of traffic at different speeds over an optical channel.

A basic DWDM system contains several main components. They are

1. **Terminal multiplexer** actually contains one wavelength converting transponder for each wavelength signal it carries. The terminal multiplexer may or may not also support a local EDFA for power amplification of the multi-wavelength optical signal.

2. **Intermediate line repeater** is placed approximately every 80–100 km for compensating the loss in optical power while the signal travels along the fiber. The signal is amplified by an EDFA which usually consists of several amplifier stages.

3. **Intermediate optical terminal** or **optical add-drop multiplexer** is a remote amplification site that amplifies the multi-wavelength signal that may have traversed up to 140 km or more before reaching the remote site.

4. **Terminal demultiplexer** breaks the multi-wavelength signal back into individual signals and outputs them on separate fibers for client-layer systems to detect.

5. **Optical Supervisory Channel** is an additional wavelength usually outside the EDFA amplification band (at 1510 nm, 1620 nm, 1310 nm). The optical supervisory channel carries information about the multi-wavelength optical signal as well as remote conditions at the optical terminal or EDFA site. It is also normally used for remote software upgrades and network operator.

DWDM systems have to maintain more stable wavelength or frequency than those needed for CWDM because of the closer spacing of the wavelengths. Precision temperature control of laser transmitter is required in DWDM systems to prevent drift of a very narrow frequency window (of the order of a few GHz). Since DWDM provides greater maximum capacity, it tends to be used at a higher level in the communications hierarchy than CWDM. Recent innovations in DWDM transport systems include pluggable and software-tunable transceiver modules capable of operating on 40 or 80 channels.

DWDM has its own limitations. Although theoretically large number of channels can be packed in a fiber, a precise wavelength selective device is needed for realization of DWDM networks. Optical amplifiers are imperative to provide long transmission distances without repeaters. However, DWDM plays an important role in high capacity optical networks and theoretically enormous capacity is possible. Practically wavelength selective (optical signal processing) components and nonlinear effects limit the performance. Passive signal processing elements like fiber Bragg grating (FBG), arrayed waveguide grating (AWG) are attractive for DWDM networks. Optical amplifications are imperative to realise DWDM networks.

4.6 OPTICAL NETWORKING

Optical network is a network involving optical components as building blocks of the network. Compared to metallic cable, fiber optic systems offer greater bandwidths, flexibility, robustness, lower attenuation, no crosstalk and no electrical interference. Those advantages have led to the dramatic growth of fiber optic systems worldwide. Today, nearly all long-haul telecommunications depend on the use of optical networks for their large capacity and robust performance.

4.6.1 Advantages of Optical Networking

1. **Size and Weight** Since individual optic fibers are typically only 125 µm in diameter, a multiple fiber cable can be made that is much smaller than corresponding metallic cables.

2. **Bandwidth** Fiber optic cables have bandwidths that can be orders of magnitude greater than metallic cable. Low-data rate systems can be easily upgraded to higher rate systems without the need to replace the fibers. Upgrading can be achieved by changing light sources (LED to laser), improving the modulation technique, improving the receiver, or using wavelength division multiplexing.

3. **Repeater spacing** With low-loss fiber optic cable, the distance between repeaters can be significantly greater than in metallic cable systems. Moreover, losses in optical fibers are independent of bandwidth, whereas with coaxial or twisted pair cable the losses increase with bandwidth. Thus, this advantage in repeater spacing increases with the system's bandwidth.

4. **Electrical isolation** Fiber optic cable is electrically non-conducting, which eliminates all electrical problems that now beset metallic cable. Fiber optic systems are immune to power surges, lightning induced currents, ground loops and short circuits. Fibers are not susceptible to electro-magnetic interference from power lines, radio signals, adjacent cable systems or other electromagnetic sources.

5. **Crosstalk** Because there is no optical coupling from one fiber to another within a cable, fiber optic systems are free from crosstalk. In metallic cable systems, by contrast, crosstalk is a common problem and is often the limiting factor in performance.

6. **Environment** Properly designed fiber optic systems are relatively unaffected by adverse temperature and moisture conditions, and therefore have application to underwater cable. For metallic cable, however, moisture is a constant problem particularly in underground (buried) applications, resulting in short circuits, increased attenuation, corrosion and increased crosstalk.

7. **Reliability** The reliability of optical fibers, optical drivers and optical receivers has reached the point where the limiting factor is usually the associated electronics circuitry.

8. **Cost** The numerous advantages listed here for fiber optic systems have resulted in dramatic growth in their application with attendant reductions in cost due to technological improvements and sales volume.

9. **Frequency allocations** Fiber (and metallic) cable systems do not require frequency allocations from an already crowded frequency spectrum. Moreover, cable systems do not have the terrain clearance, multipath fading and interference problems common to radio systems.

4.6.2 Optical Network Architecture

There are two standard optical architectures – linear and ring – both of which can provide network protection and restoration of services. **SONET** (Synchronous Optical Networking) rings are the most widely deployed architecture. The rings are designed to guarantee automatic restoration of services when cable or nodes fail by use of loops around the failed component, and hence they are called as *self-healing rings*. The **SDH** (Synchronous Digital Hierarchy) is an international standard that is highly popular and used for its high-speed data transfer of the telecommunication and digital signals. This synchronous system has been specially designed in order to provide a simple and flexible network infrastructure. This system has brought a considerable amount of change in the telecommunication networks that were based on the optical fibers as far as performance and cost were concerned.

4.6.3 SYNCHRONOUS OPTICAL NETWORKING (SONET) AND SYNCHRONOUS DIGITAL HIERARCHY (SDH)

Synchronous Optical Networking (SONET) SONET is a powerful, highly scalable technology. Although it may appear to be complex, a SONET network is transparent to the user. The Synchronous Optical network is a standard for fiber optic communication transport, and it was developed in the mid-1980s for the public telephone network. It remains in widespread use today. Synchronous Optical Networking and Synchronous Digital Hierarchy are standardised multiplexing protocols that transfer multiple digital bit streams over optical fiber using lasers or highly coherent light from light-emitting diodes. SONET commonly transmits data at speeds between 155 megabits per second (Mbps) and 2.5 gigabits per second (Gbps). Compared to Ethernet cabling that spans distances up to 100 meters, SONET fiber typically runs much further. Even short reach links span up to 2 kilometers, intermediate and long reach links cover dozens of kilometers. The increased configuration flexibility and bandwidth availability of SONET provides significant advantages over the older telecommunication systems. These advantages include the following

1. Reduction in equipment requirements and an increase in network reliability

2. Availability of set of generic standards that enable products from different vendors to be connected

3. Flexible architecture capable of accommodating future applications with a variety of transmission rates.

SONET defines optical carrier levels and electrically equivalent synchronous transport signals for the fiber optic based transmission hierarchy. SONET offers the technology for carrying many signals of different capacities through a synchronous, flexible, optical hierarchy. This is accomplished by means of a byte interleaved multiplexing scheme. Byte interleaving simplifies multiplexing and offers end to end network management. The first step in SONET multiplexing process involves the generation of the lowest level or base signal. In SONET, the base signal is referred to as synchronous transports signal-level 1(STS–1) which operates at 51.84 Mbps and SONET uses a basic transmission rate of STS

–1. Higher level signals are integer multiples of base rate. SONET's self-healing fiber optic ring functionality enables automatic network recovery due to failures that can be caused by a fiber optic cable cut, lost signal, or degraded signal (e.g. due to aging laser) or node/system failure. SONET is also a technology that is designed to ensure network traffic is restored within 60 milliseconds in the event of a failure.

SONET and SDH, which are essentially the same, were originally designed to transport circuit mode communications from a variety of different sources, but they were primarily designed to support real-time, uncompressed, circuit-switched voice encoded in PCM format. The primary difficulty in doing this prior to SONET/SDH was that the synchronization sources of these various circuits were different. This meant that each circuit was actually operating at a slightly different rate and with different phase. SONET/SDH allowed for the simultaneous transport of many different circuits of differing origin within a single framing protocol.

Due to SONET/SDH's essential protocol neutrality and transport-oriented features, SONET/SDH was the obvious choice for transporting asynchronous transfer mode (ATM) frames. It quickly evolved mapping structures and concatenated payload containers to transport ATM connections.

SONET technology enables a number of different network topologies to solve networking requirements, including survivability, cost and bandwidth efficiencies. The following provides a description of 3 different SONET configurations, which are deployed in a variety of enterprise situations. The SONET configurations include

1. Point-to-point configuration,

2. Hubbed configuration and

3. Linear Add/Drop configuration

4. Ring configuration.

Point-to-Point Configuration Point-to-point configurations are typically deployed in transport applications, which require a single SONET multiplexer in a single route. Point-to-point configurations can be enhanced to increase survivability by deploying a protection path (second fiber span) over a different path between two or more SONET multiplexers.

Hubbed Configuration Hubbed configurations consolidate traffic from multiple sites onto a single optical channel, which then can be forwarded to another site. This topology helps to reduce the number of hops as well as the equipment required to create a multisite topology.

Linear Add/Drop Configuration In the asynchronous digital signal hierarchy environment, every time a digital signal is accessed the entire signal needs to be multiplexed/demultiplexed, costing time and money at each site along a given path. However, a Linear Add/Drop configuration enables direct access to VTS/STS channels at each intermediate site along a fiber optic path. Therefore, the Linear Add/Drop configuration eliminates the need to process (multiplex/demultiplex) the entire optical signal for pass-through traffic.

Ring Configuration In a Self-Healing Ring configuration, a mechanism referred to as Automatic Protection Switching is employed. There are two types of protection ring topologies. The first is UPSR (unidirectional Path Switched Ring), the other is BLSR (Bi–directional Line Switched Ring). There are three main types of operational problems that can occur where the protection ring will take over and become the fiber span. They are

1. A break in the fiber cable

2. Signal failure (laser problem),

3. Signal degrade (due to the failure of old laser due to age) and

4. Node failure.

SONET Equipment Layers SONET defines the end-to-end connection as being made up of 3 different equipment layers, including Path Terminating Equipment (PTE), Line Terminating Equipment (LTE), and Section Terminating Equipment (STE). Figure 4.19 illustrates where each terminating equipment function resides within a SONET network.

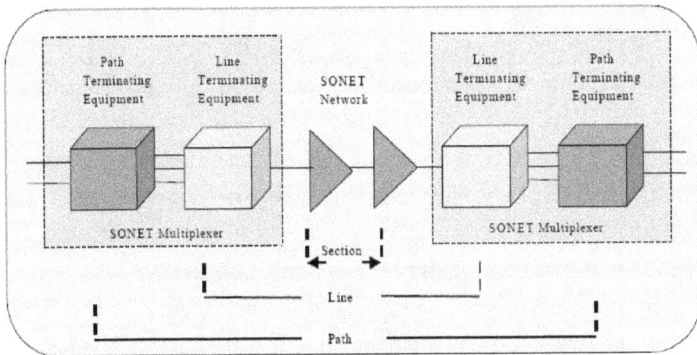

Figure 4.19 SONNET equipment layers

Protocol SONET and SDH often use different terms to describe identical features or functions. This can cause confusion and exaggerate their differences. With a few exceptions, SDH can be thought of as a superset of SONET. The protocol is a heavily-multiplexed structure, with the header interleaved between the data in a complex way.

Synchronous Digital Hierarchy (SDH) SDH (Synchronous Digital Hierarchy) is an international standard for high-speed telecommunication over optical/electrical networks which can transport digital signals in variable capacities. It is a synchronous system which intends to provide a more flexible, yet simple network infrastructure. SDH and its American version SONET emerged from standard bodies somewhere around 1990. These two standards create a revolution in the communication networks based on optical fibers, in their cost and performance.

PDH or the plesiochronous digital hierarchy is a popular technology that is widely used in the networks of telecommunication in order to transport the huge amounts of data over the digital equipment for transportation like microwave radio or fiber optic systems. The term plesiochronous has been derived from the Greek work 'plesio' that means 'near' and 'chronos' meaning time. This means that the PDH works in a state when the various different parts of a network are clearly synchronized. But with the change in technology, the PDH is now being replaced by the SDH or what is popularly called as synchronous digital hierarchy. SDH is a standard technology for synchronous data transmission on optical media. It is the international equivalent of Synchronous Optical Network. Both technologies provide faster and less expensive network interconnection than traditional PDH (Plesiochronous Digital Hierarchy) equipment. This PDH helps in proper transmission of the data that generally runs at the similar rate but allows some slight variation in the speed than the nominal rate. The basic transfer rate of the data is 2048 kilobits per second. For instance, in each speech transmission, the normal rate breaks into different thirty channels of 64 kilobits per second along with two different 64 kilobits per second in order to perform the tasks of synchronization and signaling. The typical rate of transmitting the data over the fiber optic systems is 565 Mbit/sec in order to transport the data in the long distance. But as the technology has improved with the passing of time, now the telecommunication companies have replaced the PDH equipment with that of the SDH equipment, which has the capability of transmitting the data at much higher rates as compared to the PDH system. The weaknesses that PDH faced paved way for the introduction and use of the SDH systems. Although the PDH proved to be a breakthrough in the field of digital transmission, the weaknesses that made it less demanded includes

1. Asynchronous structure that is rigid,

2. Restricted management capacity,

3. Non availability of world standard on the digital formats and

4. No optical interfaces world standard and without an optical level, networking is not possible.

In digital telephone transmission, synchronous means the bits from one call are carried within one transmission frame. Plesiochronous means 'almost (but not) synchronous' or a call that must be extracted from more than one transmission frame. SDH uses the following Synchronous Transport Modules (STM) and rates – STM–1 (155 megabits per second), STM–4 (622 Mbps), STM–16 (2.5 gigabits per second), and STM–64 (10 Gbps).

SDH is very useful equipment that is used for the telecommunication sector in easy transfer of data. Earlier PDH was widely used but due to some of its weaknesses, SDH has replaced the use of PDH. Point-to-Point applications are still used mainly by PDH as it is cheaper. But if one compares the PDH system with that of the SDH system, the latter one has a large number of advantages. Some of the most common advantages enjoyed by the usage of SDH include

1. optical interfaces,

2. capability of powerful management,

3. world standard digital format,

4. synchronous structure is flexible,

5. cost effective and easy traffic cross connection capacity and add and drop facility,

6. reduced networking cost due to the transversal compatibility and

7. forward and backward compatibility.

Apart from all the advantages mentioned above, the SDH also has various management capabilities such as performance management, security and access management, configuration management and the event or the alarm management. Therefore, we can clearly make a distinction between the PDH and SDH systems so that as per the needs of the telecommunication, the appropriate transmission system can be used.

4.7 ACTIVE AND PASSIVE OPTICAL NETWORKS

Fiber optics uses light signals to transmit data. As this data moves across a fiber, there needs to be a way to separate it so that it gets to the proper destination. There are two important types of systems that make fiber-to-the-home broadband connections possible. These are active optical networks and passive optical networks and each has their own advantages and disadvantages.

4.7.1 Active Optical Networks (AON)

An active optical system uses electrically powered switching equipment, such as a router or a switch aggregator, to manage signal distribution and direct signals to specific customers. This switch opens and closes in various ways to direct the incoming and outgoing signals to the proper place. In such a system, a customer may have a dedicated fiber running to his or her house. Active optical network has many advantages such as point-to-point, high bandwidth and high flexibility. Their reliance on Ethernet technology makes interoperability among vendors easy. Subscribers can select hardware that delivers an appropriate data transmission rate and scale up as their needs increase without having to restructure the network. Active optical networks, however, also have their weaknesses. They require at least one switch aggregator for every 48 subscribers. Because it requires power, an active optical network inherently is less reliable than a passive optical network.

Active optical networks offer certain advantages, as well. Their reliance on Ethernet technology makes interoperability among vendors easy. Subscribers can select hardware that delivers an appropriate data transmission rate and scale up as their needs increase without having to restructure the network. Active optical networks, however, also have their weaknesses. They require at least one switch aggregator for every 48 subscribers. Because it requires power, an active optical network inherently is less reliable than a passive optical network.

4.7.2 Passive Optical Networks (PON)

A Passive Optical Network (PON) is a single, shared optical fiber that uses inexpensive optical splitters to divide the single fiber into separate strands feeding individual subscribers. PONs are called 'Passive' as there are no active electronics within the access network except subscriber points.

A passive optical network does not include electrically powered switching equipment and instead uses optical splitters to separate and collect optical signals as they move through the network. A passive optical network shares fiber optic strands for portions of the network. Powered equipment is required only at the source and receiving ends of the signal. The advantage of passive optical networks is that they are efficient in such a way that each fiber optic strand can serve up to 32 users. PONs have a low building cost relative to active optical networks along with lower maintenance costs. As PON has few moving or electrical parts, going wrong in a PON is least. On the other hand, passive optical networks also have some disadvantages. They have less range than an active optical network. PONs also makes it difficult to isolate a failure when they occur. Since the bandwidth in a PON is not dedicated to individual subscribers, data transmission speed may slow down during peak usage times (latency). Latency quickly degrades services such as audio and video, which need a smooth rate to maintain quality.

A passive optical network (PON) is a system that brings optical fiber cabling and signals all or most of the way to the end user. Depending on where the PON terminates, the system can be described as fiber-to-the-curb (FTTC), fiber-to-the-building

(FTTB), or fiber-to-the-home (FTTH). A passive optical network (PON) is a point–to–multipoint, fiber to the premises network architecture in which unpowered optical splitters are used to enable a single optical fiber to serve multiple premises. A PON configuration reduces the amount of fiber and central office equipment required compared with point-to-point architectures.

4.7.3 Fiber to the Home Broadband Connections

In some cases, FTTH systems may combine elements of both passive and active architectures to form a hybrid system. A key benefit to FTTH (also called FTTP) is that it provides for far faster connection speeds and carrying capacity than twisted pair conductors, DSL or coaxial cable. FTTH (fiber to the home) is a form of fiber optic communication delivery that reaches one living or working space. The fiber extends from the central office to the subscriber's living or working space. Once at the subscriber's living or working space, the signal may be conveyed throughout the space using any means, including twisted pair, coaxial cable, wireless, power line communication or optical fiber.

FTTH will be able to handle even the futuristic Internet uses. Advanced technologies such as 3D holographic high definition television and games will someday be everyday items in households around the world. FTTH will be able to handle the estimated 30-gigabyte-per-second needs of such equipment. Current technologies can't come close. The FTTH broadband connection will spark the creation of products not yet dreamed of as they open new possibilities for data transmission rate. FTTH broadband connections also will allow consumers to take advantage of multipurpose communications services. A consumer could receive telephone, video, audio, television and just about any other kind of digital data stream using a simple FTTH cost effective broadband connection

REVIEW QUESTIONS

1. What are the importances of optical amplifiers?
2. Explain the basics of an optical amplifier.
3. What are the types of optical amplifiers?
4. Describe a semiconductor optic amplifier and its characteristics.
5. List the advantages and disadvantages of EDFA.
6. With energy level diagram of Eb, explain the working of a EDFA.
7. Describe EDFA amplifier configuration with necessary diagram and its characteristics.
8. How will you accomplish the gain equailisation in rare earth doped optical amplifier?
9. Fiber amplifiers gain signal and not noise. Comment.
10. Write a note on erbium in C-band.
11. What are the properties of Raman amplifiers?
12. List the difficulties with Raman amplifiers.
13. With necessary diagram, describe the Raman amplifier. List its salient features.
14. What are optical filters? Mention their uses in optical communication.
15. Why single mode fibers are the best choice for long distance communications?
16. Explain optical attenuation and their uses.
17. Write notes on
    i. circulators,

    ii.   isolators and

    iii. optical switches.

18. List the common applications of couplers and splitters. Mention their applications.

19. What is wavelength converter? What are its functions?

20. What are the characteristics of an ideal wavelength converter?

21. With necessary diagram, describe and explain optoelectroxnic wavelength converter.

22. Describe a fiber optic transceiver.

23. What are the functions of link design? Discuss a typical optical link power budget.

24. Distinguish between WDM and DWDM.

25. Why WDM is necessary?

26. What are the constraints in WDM networks?

27. List the components in DWDM

28. What are the advantages of optical networking?

29. Discuss an optical network architecture explaining SONNET and SDH

30. Write notes on (a) AON, (b) PON and (c) FTTH

CHAPTER-V

OPTOELECTRONIC INTEGRATED CIRCUITS

5.1 INTRODUCTION

Optoelectronic integration is nothing but integrating optoelectronic devices (such as lasers and photodiodes) and electronic devices together. This emerging technology is a boost in developing future lightwave systems. The first experiment demonstrating optoelectronic integrated circuit (OEIC) was done in 1978–79 by integrating GaAs-based lasers and driver diodes and transistors. This technology has its potential advantages in realizing high-performance, high-manufacturability and high-functionality in optoelectronic components which have not been developed fully due to the limit of using conventional discrete devices and assembly technique. Optoelectronic integrated transmitters and receivers are currently being developed for lightwave telecommunication and interconnection applications by a number of researchers, and recent progresses in both the performance and the fabrication technology are encouraging. Technologies for integrating optoelectronic devices and electronic circuitry can be classified as either hybrid or monolithic. None of these technologies is fully developed but limited hybrid integration is available commercially. Hybrid integration involves combining optoelectronic devices and integrated circuits in the same package or substrate. This can be done using traditional hybrid techniques for simply combining packaged devices on a ceramic substrate, but the device density of the resulting OEIC is very low and many of the advantages of using optics are lost. A variety of integrated optoelectronic devices can be fabricated from a simple semiconductor laser diode waveguide process. Technological challenges in optoelectronic integration will lead to a generation of future advanced optical systems for not only communication but also switching and computing. Integrated optical amplifiers have the advantage of integrating with other guided wave devices such as power dividers leading to compensation of waveguide losses. And also optical switches associated with integrated optical amplifiers are capable of providing loss less switching and it will reduce the cross talk. Integrated optical amplifiers are much smaller in length than fiber amplifiers.

Not only have OEIC transmitters and receivers been developed for telecommunications, but also the novel optical functional devices and photonic integrated circuits have been investigated and developed for advanced telecommunication networks and data processing and sensing systems.

Optoelectronic integrated circuit (OEIC) has often been used to indicate the monolithic integration of optical and electronic devices on a single common substrate. The materials most often used in optoelectronic integration are GaAs- and InP-based III–V semiconductor alloy systems.

5.2 HYBRID AND MONOLITHIC INTEGRATION

Hybrid and monolithic integration technology is used in fabricating optoelectronic integrated circuits. Commonly used materials along with integration technology, wavelength and features are listed in Table 5.1.

Table 5.1 Materials and its features for optoelectronic integrated circuits (courtesy: Osamu Wada, Optoelectronic Integrated Circuits)

Integration technology	Processing technique	Substrate materials	Heterostructure materials	Wavelength	Features
Monolithic	Lattice matched epitaxy	GaAs	AlGaAs/GaAs	0.8 µm	Short distance transmission, Advanced electronic LSIs.
	Lattice matched epitaxy	InP	InGaAsP/InP, InGaAlAs/InP	1.3–1.55 µm	Long distance transmission, Very fast electronic ICs.
	Lattice mismatched heteroepitaxy	Si, GaAs, InP	GaAs- and InP-based	0.8–1.55 µm	Low-defect density growth required, Potential use of Si LSIs.

Table 5.1 *(Continued)*

Integration technology	Processing technique	Substrate materials	Heterostructure materials	Wavelength	Features
	Standard Si bipolar and CMOS technology	Si	All Si	visible, 0.8μm	No light emitting devices available, Advanced LSIs, Low-cost chips.
Hybrid	Direct wafer bonding	Semiconductors, dielectrics	GaAs- and InP-based, etc.	0.8 μm–1.55 μm	Reliable wafer bonding technique required, integratable with Si LSIs.
	Flip-chip bonding	Semiconductors, dielectrics	GaAs- and InP-based, etc.	0.8–1.55 μm	Bonding required after wafer processes, Integratable with Si LSIs.

5.2.1 Hybrid Circuits

Hybrid integration involves combining optoelectronic devices and integrated circuits in the same package or substrate. This can be done using traditional hybrid techniques for simply combining packaged devices on a ceramic substrate, but the device density of the resulting optoelectronic integrated circuit (OEIC) is very low and many of the advantages of using optics are lost. Consequently, alternative hybrid techniques have been proposed to increase device density while maintaining manufacturability. Currently, the most successful such techniques involve flip-chip/solder-ball or -bump integration of discrete optoelectronic devices on multi-chip modules or directly on silicon IC chips, but even these techniques consume an inordinate amount of area and even higher densities are needed. A possible solution is the use of specially thinned discrete devices (and, ultimately, moderate sized arrays of devices) which can be flip-bonded to IC chips with much smaller bond pads, and thus with more efficient use of space (and even some stacking of devices). Prof. N. Jokerst and her students at Georgia Tech have done the most work developing this technology, which they term "epitaxial lift-off". This technique was first demonstrated by E. Yablonovich at BellCore (now at UCLA).

Hybrid integration offers an immediate solution to certain OEIC needs and it may always be the technology of choice when many different kinds of devices need to be combined, but monolithic integration will always be superior in terms of speed, device density, system reliability, ultimate complexity and manufacturability. A major difficulty with monolithic integration is that the most widely used, highly developed material for integrated electronics, silicon, is not useful for many optoelectronic devices and thus one must either monolithically integrate III–V optoelectronic devices on silicon, or one must utilize III–V electronics.

Figure 5.1   An (orange-epoxy) encapsulated hybrid circuit on a printed circuit board (Courtesy: Wikipedia)

A hybrid integrated circuit (HIC), hybrid microcircuit, or simply hybrid is a miniaturized electronic circuit constructed of individual devices, such as semiconductor devices (e.g. transistors and diodes) and passive components (e.g. resistors, inductors, transformers, and capacitors), bonded to a substrate or printed circuit board (PCB). Hybrid circuits are often encapsulated in epoxy as shown in Figure 5.1. A hybrid circuit serves as a component on a PCB in the same way as a monolithic integrated circuit. The difference between the two types of devices is in the technique in construction and manufacturing. The advantage of hybrid circuits is that components which cannot be included in a monolithic IC can be used, e.g., capacitors of large value, wound components, crystals.

Thick film technology is often used as the interconnecting medium for hybrid integrated circuits. The use of screen printed thick film interconnect provides advantages of versatility over thin film although feature sizes may be larger and deposited resistors wider in tolerance. Multi-layer thick film is a technique for further improvements in integration using a screen-printed insulating dielectric to ensure connections between layers are made only where required. One key advantage for the circuit designer is complete freedom in the choice of resistor value in thick film technology. Planar resistors are also screen printed and included in the thick film interconnect design. The composition and dimensions of resistors can be selected to provide desired values. The final resistor value is determined by design and can be adjusted by laser trimming. Once the hybrid circuit is fully populated with components, fine tuning prior to final test may be achieved by active laser trimming.

5.2.2 Monolithic Integration

Monolithic integration of InGaAlAs and InGaAsP heterostructures on silicon (a process generically called GaAs-on-Si) is attractive because it would allow one to make use of the wealth of silicon BiCMOS electronics integrated circuit technology existing today. Unfortunately, III–V epitaxy on silicon (or vice versa) is made difficult by the facts that the lattice constants of GaAs and Si differ by 4%, and that the thermal expansion coefficients differ by almost 50%. The lattice mismatch can be overcome by special growth initiation techniques and interfacial buffer layers, but the thermal mismatch is more troublesome. Significant numbers of defects (and even cracks) are introduced as GaAs-on-Si structures are cooled from the growth temperature to room temperature and the III–V layers are under severe tension. This reduces device lifetimes dramatically, reduces reliability, and adversely affects performance. The most promising solution to this incompatibility involves introducing an AlAs layer between the growth-initiation layers and the active device heterostructures and subsequently AlAs layer is etched away to separate the III–V device structure physically from the Si. Preliminary studies at MIT have shown that the separated structure can be thermally annealed to remove the thermally induced strain and defects (significantly improving the material quality and device performance) but much more work remains to be done.

The difficulty of monolithic integration on silicon and steady advances in the levels of electronic integration demonstrated on III–V substrates have led to a shift of emphasis to monolithic integration on either gallium arsenide or indium phosphide.

Two approaches are being adopted. The 'epitaxy-first' or 'stacked-heterostructure' methods involve first growing all of the epitaxial layers needed for both the optoelectronic devices and electronic devices on a suitable substrate. The top layers are then removed selectively in areas where devices are to be fabricated on lower layers and each of the various device types are formed in the appropriate layers. On the other hand, some of the epitaxial layers are grown and unwanted layers are etched away. Subsequently, more epitaxial layers are grown in newly cleared areas. This approach results in a more planar surface but the growth and processing are much more complex. An important problem with the epitaxy-first approach is that pursuing it involves simultaneous development of sophisticated technologies for optoelectronic devices and integrated circuits. The task is enormous and as practical matters only small scale OEICs have been demonstrated following this approach.

While the integration levels of epitaxy-first III–V OEICs have remained low, the degree of integration in commercial GaAs electronic (MESFET) integrated circuits has reached VLSI levels in recent years and the technologies used in this integration use only refractory metals in contact with the semiconductor and use Si IC upper-level interconnect and passivation technologies. These advances offer a route to achieving much higher levels of optoelectronic integration through epitaxial growth of III–V heterostructures on GaAs-based VLSI electronics. This is essentially the GaAs-on-Si idea without the myriad of materials problems inherent in growing the III–Vs on silicon and with all the advantages of building on a commercial integrated electronics foundation. This approach is called epitaxy-on-electronics (epi-on-electronics) and gives high density and high performance optoelectronic integrated circuits.

Optoelectronic integrated circuits are applied in computation, parallel processing of data and images, en/decryption using Smart Pixel Arrays. They are also used in diffuse optical tomography for seeing tumors, blood vessels, bones, etc. beneath the skin. Here OEIC chip is fabricated with interwoven arrays of detectors and emitters. In this chip, each VCSEL is illuminated in turn and the pattern of scattered light seen by the detector array is recorded. With this information, an image of the sub-surface structure can be constructed. The near infrared light is strongly scattered but weakly absorbed in soft body tissue.

5.3 OE INTEGRATED TRANSMITTERS AND RECEIVERS

Transmitters and receivers consist of optical and electronic devices having very different structures and fabrication processes. Most work in this regard has gone into attaining process compatibility and developing reproducible fabrication techniques. Osamu Wada has given a detailed account of integrated transmitters and receivers in Handbook of Electro-Optics, (Ed. R. Waynant) McGraw-Hill, New York, 2000 in chapter 27.

5.3.1 Transmitters

In transmitter fabrication, low-threshold current lasers are important in minimizing heating and hence Quantum well lasers are extensively used to lower the threshold current. Transverse mode-stabilized laser structures such as BH and ridge-waveguide structures are used to ensure stable lasing. Figure 5.2 shows (*a*) the structure and (*b*) the circuit diagram of a multichannel OEIC transmitter array fabricated on a Si-GaAs substrate. MBE-grown AlGaAs/GaAs GRINSCH single-quantum-well (SQW) lasers having ridge waveguide structures are planar embedded in a Si-GaAs substrate and the lasers inner facets are formed by micro cleavage technique. Laser power monitor photodiodes are formed by etching the same heterostructure. The driver circuit consists of three GaAs MESFETs. In InP-based materials systems, simple transmitters involving a cleaved-facet laser and a single to a few driver transistors have been reported in the literature.

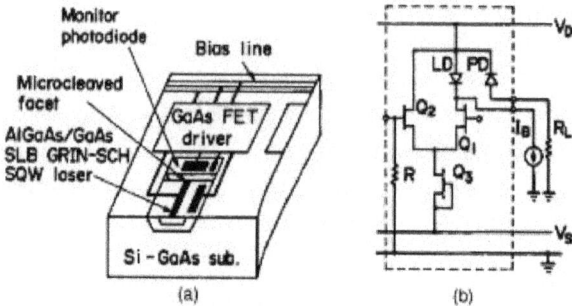

Figure 5.2   (a) structure and (b) circuit diagram of multichannel GaAs-based OEIC transmitter array containing QW laser. Monitor photodiodes and GaAs MESFET driver circuits.

Lattice-mismatched heteroepitaxy and wafer bonding are important to combine materials for enhancing the advantages in integration. A transmitter consisting of a GaInAsP/InP BH laser and GaAs MESFETs has been fabricated by heteroepitaxy and excellent transmission characteristics at 1.2 Gb/s has been reported in the literature. Laser transmitter fabrications on Si substrates have not yet been achieved but AlGaAs/GaAs LED/Si MOSFET integration and single GaAs-based laser fabrication on Si have already been demonstrated.

5.3.2 Receivers

Simple receivers consisting of p-i-n photodiodes and preamplifiers have been reported in the literature using vertical and horizontal integration structures. Planar compatible structures incorporating a MSM photodiode and MESFETs on an Si-GaAs have been reported and this will increase the integration scale without compromising manufacturability. Figure 5.3 shows

(a) the structure and

(b) the circuit diagram for an eight-channel MSM/amplifier receiver array involving 208 devices.

An MSM photodiode 100 μm^2 in area and with 3 μm lines and spaces for interdigital electrodes has been integrated with

2-μm gate MESFETs to construct a transimpedance amplifier. Overall receiver sensitivity is uniform, within 105 V/W ±10 V/W over an eight-channel array, confirming the advantage of planar, compatible process. MSM photodiodes also help achieve low-capacitance, high-speed, and low-noise performance. The cutoff frequency of 1.1 GHz and the noise floor of 5 pA/Hz$^{\frac{1}{2}}$ are consistent with the circuit parameters used. The structure and the circuit diagram of GaAs MSM/amplifier OEIC receiver is shown in Figure 5.3. The crosstalk is extremely low, less than –37 dB up to 0.6 GHz. Essentially, the same structures as used for MSM photodiodes have been used in many applications for large-scale OEIC receivers.

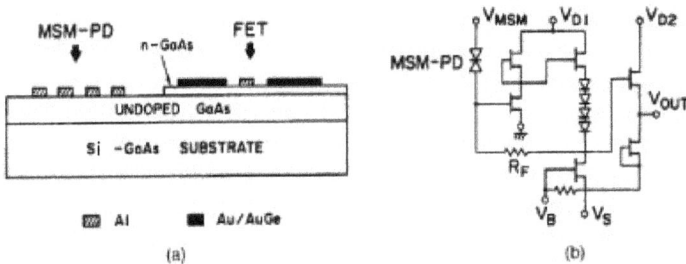

Figure 5.3 (a) Structure and (b) Circuit diagram of GaAs MSM/amplifier OEIC receiver

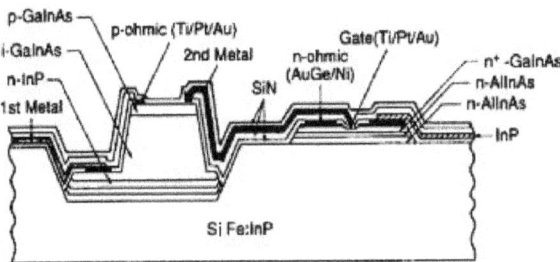

Figure 5.4 Cross sectional view of PIN/HEMT monolithic OEIC receiver (After Yano, et al.1992)

Since early reports on simple PIN/FET OEICs, InP-based OEIC receivers have been developed most extensively among all

OEICs due to their practical importance in optical communication systems. Yano, Sasaki, Murata and Hayashi reported the structure of a PIN/HEMT preamplifier OEIC receiver in 1992 in IEEE Tran. Electron Devices, 39, 2254. Figure 5.4 shows the structure of a PIN/HEMT preamplifier OEIC receiver.

A single epitaxial growth on a patterned InP substrate has been applied to realize a planar, compatible OEIC structure to integrate an AlInAs-capped MSM photodiode and InAlAs/InGaAs HEMTs as shown in Figure 5.5.

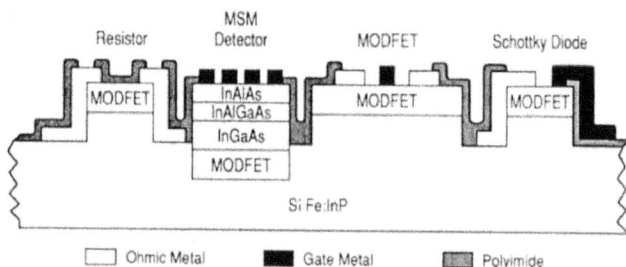

Figure 5.5 Cross section of MSM/MODFET monolithic OEIC receiver with quasi planar embedded structure (After Hong et al. 1989)

There has been excellent progress in the field of HBT-based receiver OEICs reaching 40 Gb/s and performance comparable to the best hybrid receivers. Chandrasekhar, et al reported PIN/HBT receiver in 1988 and in 1995. Figure 5.6 shows the structure (a) and the circuit (b) of a PIN/HBT receiver.

The amplifier design is a double feedback input stage having an increased bandwidth with high gain using double heterojunction devices (DHBTs). This circuit has operated up to 20 Gb/s with a sensitivity of −17 dBm for a bit error rate (BER) of 10^{-9}. More recently, the operation of PIN/HBT OEIC receivers has been demonstrated at 40 Gb/s. Lattice-mismatched heteroepitaxy has been used to combine a GaInAs/InP PIN photodiode and a GaAs MESFET amplifier on an InP substrate. However, hybrid integration using flip-chip bonding has been developed extensively for practical applications to receiver integration.

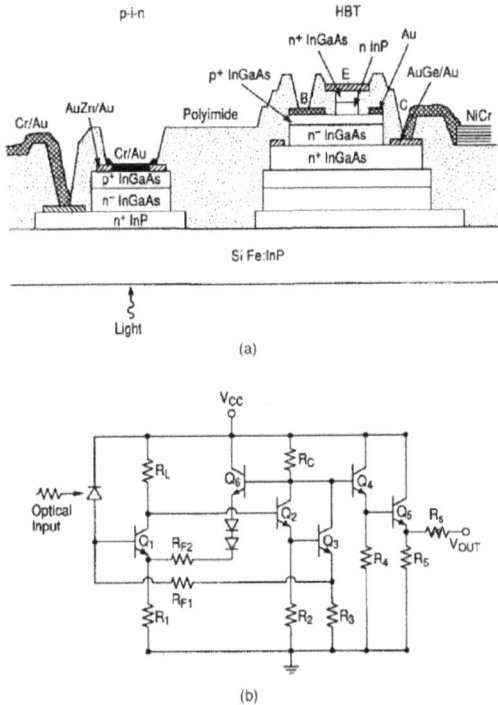

Figure 5.6   (a) Cross section, (b) Circuit diagram of PIN/HBT monolithic OEIC receiver (After Chandrasekhar et al. 1988, 1995)

Si-based OEIC Receivers   The indirect gap of this material makes it difficult to use for light emission, but it is now possible by using the recent invention of luminescence from porous Si. However, this material has been used in OEIC receivers in the 0.8 μm wavelength band, waveguide circuits and substrates for hybrid integration of various optical components. Si-based PIN/ bipolar IC receiver was reported for short distance optical link applications. Although Si PIN photodiode is limited in speed due to small absorption coefficient –10 to –15 dBm, sensitivity has been observed at 125–500 Mb/s data rate. An OEIC receiver with simpler bipolar structures was developed by H. Nagao and

M. Yamamoto in 1989, and it is shown in Figure 5.6. They have shown the performance up to approximately 50 Mb/s. Such Si-based OEIC receivers have been used in office data links as well as in optical sensing and optical storage areas, as being represented by optical pick-up in audio disk and minidisk products.

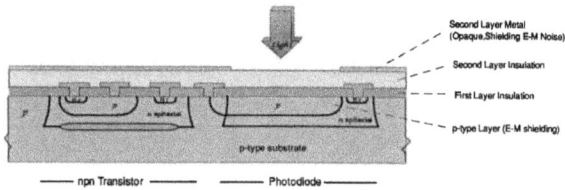

Figure 5.7 Si-based OEIC incorporating a p-n photodiode and bipolar transistor signal processing circuit (After Nagao et al. 1989)

REVIEW QUESTIONS

1. Distinguish between hybrid and monolithic integrated technology.

2. What is a hybrid circuit? Explain.

3. What are the advantages of hybrid circuits?

4. Discuss any two methods for monolithic integration in GaAs or InP.

5. Explain in detail about the fabrication and uses of monolithic integrated circuits.

6. Describe the structure and circuit diagram of multichannel GaAs based OEIC transmitter.

7. With necessary diagram, explain the GaAs MSM/amplifier OEIC receiver.

8. Draw the cross sectional view of PIN/HEMT monolithic OEIC receiver.

9. Draw and explain the cross sectional view of MSM/MODFET monolithic OEIC receiver embedded with quasi planar structure.

10. Draw and explain the PIN/HBT monolithic OEIC receiver.

11. Write a note on Si-based OEIC receiver.

Chapter – VI

APPLICATION OF OPTICAL FIBERS

6.1 FIBER OPTIC SENSOR

A fiber optic sensor is a sensor that uses optical fiber as the sensing element. Fiber optic sensors are often loosely grouped into three basic classes referred to as intrinsic or all fiber sensors (fiber itself is the transducer), extrinsic or hybrid fiber optic sensors (transducer acts on the fiber) and hybrid sensors (fiber carries light in and out of the device). The intrinsic fiber optic sensor that has a sensing region within the fiber and light never goes out of the fiber. They measure temperature, pressure, refractive index, strain, displacement, force and load using the Fabry-Perot principle. On the other hand, in extrinsic sensors light leaves the fiber and reach the sensing region outside and then comes back to the fiber. Figure 6.1 illustrates classification of optical fiber sensors.

Compared with other types of sensors, fiber-optic sensors exhibit a number of advantages. They are

1. They are extremely accurate with very high precision and absolute measurement.

2. They are intrinsically safe and passive operation.

3. They are minimally invasive.

4. They consist of electrically insulating materials (no electric cables are required) and hence they can be used in high-voltage environments.

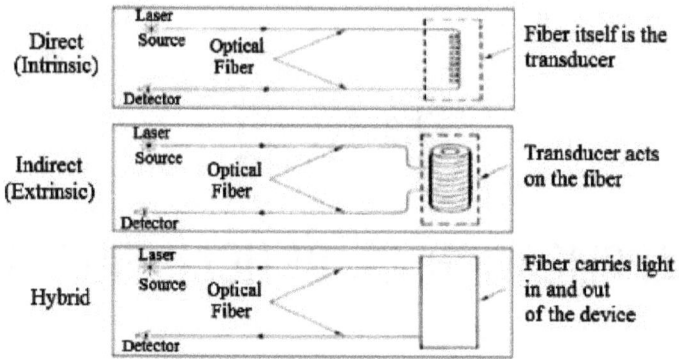

Figure 6.1 Classification of optical fiber sensors (Courtesy: Optical fiber sensors guide, Micron Optics)

5. They can be safely used in explosive environments as there is no risk of electrical sparks, even in the case of defects.

6. They are immune to electromagnetic interference (EMI), radio frequency and microwave radiation.

7. Their materials are chemically passive. They do not contaminate their surroundings and water and corrosion resistant.

8. They have a very wide operating temperature range.

9. They have multiplexing capabilities. Multiple sensors in a single fiber line can be interrogated with a single optical source.

10. They possess light weight, rugged and small size.

11. They offer excellent performance with high sensitivity and large bandwidth.

12. Multiplexed in parallel or in series or distributed measurements can be done with them.

13. They can be used in harsh environment capability such as intensive EMI, high temperature, chemical corrosion, high pressure and high voltage.

14. They give excellent resolution and range.

15. They have modest cost per channel.

Fiber optic sensors exhibit shortcomings too. Sensors with immobilized dyes and other indicators have limited long-term stability and their shelf life degrades over time. Further, ambient light can interfere with the optical measurement unless optical shielding or special time-synchronous gating is performed.

Fabry–Perot sensors are widely used in different applications in medical, aerospace and defense, energy, scientific and process control. Extrinsic fiber optic sensors use an optical fiber cable to transmit modulated light from either a non-fiber optical sensor or an electronic sensor connected to an optical transmitter. This type of sensors reaches places that are inaccessible. For example, temperature measurement inside aircraft jet engines and internal temperature of electrical transformers etc. Fiber optic sensors are ideal for harsh conditions, including high vibration, extreme heat, noisy, wet, corrosive or explosive environments. Fiber optic sensors are small enough to fit in confined areas and can be positioned precisely.

Sensors used can be either standard or disposable. Disposable fiber optic sensors are used to meet regulatory requirements such as

a. cleaning validation,

b. to avoid the spread of infection and

c. cross-contamination in equipment or to operators.

Fiber optic sensors measure In medical applications it is used in

a. patient monitoring in MRI,

b. intravascular blood pressure monitoring,

c. intracranial pressure monitoring,

d. intrauterine pressure monitoring and

e. cardiac guide wire-mounted pressure monitoring.

In aerospace and defense applications, it is used in high-speed temperature characterization for electro-explosive devices (EED). In process control applications it is used to measure refractive index, temperature, oil spilling, pressure and strain measurement for industrial environment. In energy sector, it is used to monitor hot spot winding temperature of electrical power transformer

Fiber optic pressure sensors are suitable for various applications such as oil and gas pumping station, plastic injection moulding and extrusion monitoring, MRI and microwave, aerospace, medical device components monitoring and in-situ process monitoring.

Fiber optic refractive index Sensors are suitable for various applications such as quality control, research and development and chemical environments.

Fiber optic strain sensors are Suitable for various applications such as research and development, structural health monitoring, nuclear power plants, corrosive environments and tunnel linings. It is resistant to corrosive environments and miniature. The advantages of Fiber optic strain sensors are

a. high sensitivity,

b. high resolution,

c. no interference due to cable bending and

d. absolute measurements in engineering units.

Fiber optic temperature sensors are suitable for various applications such as microwave, RF, in-situ process monitoring, nuclear, medial and aerospace.

6.2 FIBER OPTIC SENSOR SYSTEMS

Fiber optics for sensing applications are used to communicate with a sensor device or use a fiber as the sensor itself to conduct continuous monitoring of physical, chemical, and biological changes in the subject or object of study.

In fiber-optic sensors, information is primarily conveyed in all optical sensors by a change in either phase, polarization, frequency, intensity or a combination. But the photo detector, being a semiconductor device, only senses intensity of light at the detector surface. Therefore, the art of sensing with polarization, phase or frequency modulation involves interferometric or grating based signal processing optical circuits. Instrumentation is designed depending on the type of optical sensor used and its intended application and they vary greatly. The block diagram of a fiber optic sensor design is shown in Figure 6.2. However, the basic fiber optic sensor design is given in Figure 6.3.

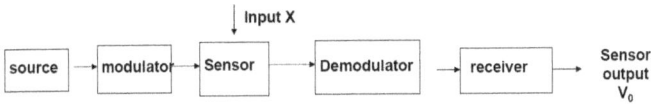

Figure 6.2 Block diagram of a fiber optic sensor design

Optical fibers are also attractive not only for sensing but also in control and instrumentation. Optical fibers have made a significant impact in these areas too. For these applications, fibers are made more susceptible and sensitive to the same external mechanisms against which fibers were made to be immune for their effective operation in telecommunications.

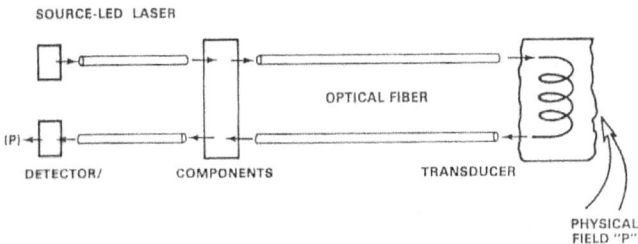

Figure 6.3 A typical optical fiber sensing system (Courtesy: Optical fiber sensors guide, Micron Optics)

A typical optical fiber sensing system is basically composed of a light source, optical fiber; a sensing element or transducer and a detector (Figure 6.3). The principle of operation of a fiber sensor is that the transducer modulates some parameter of the optical system (intensity, wavelength, polarization, phase, etc.) which gives rise to a change in the characteristics of the optical signal received at the detector.

6.3 APPLICATIONS OF FIBER OPTIC SENSORS (FOS)

Fiber optic sensing technologies have formed an entirely new generation of sensors offering many important measurement opportunities and great potential for diverse applications. The most highlighted application fields of FOS are in large composite and concrete structures, the electrical power industry, medicine, chemical sensing, and the gas and oil industry.

FOS technology offers greater resistance to corrosion when used in open structures, such as bridges and dams. It is used for quality control during construction, health monitoring after building, impact monitoring of large composite or concrete structures. Since composite materials are not well proven in their substitution for steel in concrete structures, there is considerable interest in monitoring the strain and deformation or deflection, temperature, or environmental degradation within such types of composite structures using an integrated fiber optic sensing system. FBG sensors could be suitable for achieving such a goal.

Dams are probably the biggest structures in civil engineering; hence it is vital to monitor their mechanical properties during and after construction in order to ensure the construction quality, longevity, and safety of the dam. Fiber optic sensors are ideal for health monitoring applications of dams due to their excellent ability to realize long-range measurement. Truly distributed FOS is particularly attractive as they normally have tens of km measurement range with meter spatial resolution. One of the most important parameter for monitoring in dams is temperature. This monitoring is of prime importance as the density and micro cracks are directly related to the maximum temperature the concrete experiences during the setting chemical process. A distributed temperature sensor using Brillouin scattering can be used to monitor this feature [Joan R. Casas, Paulo J.S. Cruz, M. Asce, 2003].

Measurement of load and displacement changes in underground excavations of mines and tunnels is most important for safety monitoring. Multiplexed FBG sensors are for load measurement and displacement changes at these places [F. Yu, S. Yin, *Fiber optic sensors*. 2002].

Advanced composite materials are now routinely used for manufacturing engineering structures such as aerospace structures (e.g., parts of airplane wings). Compared with metallic materials, advanced composite materials can have higher fatigue resistance, lighter weight, higher strength to weight ratio, the capability of obtaining complex shapes and no corrosion. Hence, the use of composite materials with embedded FBG systems can lead to a reduction in weight, inspection intervals, and maintenance cost of aircraft and consequently to an improvement in performance. FBG sensors are sensitive to both strain and temperature. Hence, it is essential to measure strain and temperature simultaneously in order to correct the thermally induced strain for static strain measurement. A simple and effective method often used is to employ an unstrained temperature reference FBG [F. Yu, S. Yin, *Fiber optic sensors*. 2002].

The FBG is ideal for use in the electrical power industry due to its immunity to electromagnetic interference. Loading of power transmission lines, winding temperature of electrical power transformers have been measured with the FBG sensor. An excessive mechanical load on electrical power transmission lines such as heavy snow in mountainous areas, Causes a serious accident. Since there is no access for inspection in these areas, an online measurement system is needed to monitor the changing load on the power line. A multiplexed FBG sensor is used here [F. Yu, S. Yin, *Fiber optic sensors*. 2002]. Knowledge of the local temperature distribution present in high-voltage, high-power equipment such as generators and transformers is essential in understanding their operation. Defective or degraded equipment can be detected by continuously monitoring the variations in the winding temperature and FBG sensors are the best choice for this application.

The majority of commercial sensors widely used in medicine are electrically active and hence not appropriate for use in a number of medical applications. Fiber optic sensors can overcome these problems, as they are virtually dielectric. A range of miniature fiber optic sensors based on intensity modulation has been successfully commercialized in recent years.

The FBG sensor can also be used for the measurement of the heart's efficiency based on the flow-directed thermo dilution catheter method in which doctors inject patients with a cold solution to measure their heart's blood output. A flow directed thermo dilution catheter is inserted into the right atrium of the heart, allowing the solution to be injected directly into the heart for measurement of the temperature of the blood in the pulmonary artery. By combining temperature readings with pulse rate, doctors can determine how much blood the heart pumps. Such a type of catheter with an FBG sensor has been used for replacement of a conventional catheter [F. Yu, S. Yin, *Fiber optic sensors*, 2002].

Fiber optic sensors could be ideal for applications in the oil and gas industry due to their inherent advantages such as being intrinsically safe, immune to EMI, workable at high temperature, capable of multiplexing and minimally invasive. Of these advantages, the multiplexed or distributed sensing feature is particularly attractive for applications where monitoring of a parameter or parameters at many spatial locations through the well is essentially necessary. FBG sensors can be used in pipeline in order to monitor the temperature in the pipe or the pressure on the joints. Because refractive index of these fibers is sensitive to pressure and temperature, one can monitor all the length of the pipe by one fiber by using multiplexing methods.

Lastly, crack detection is very important in buildings, especially in concrete bridges and dams. The crack openings beyond 0.15 to 0.2 mm will allow the penetration of water and chloride ions leading to the corrosion of steel reinforcements. Generally, crack detection and monitoring for bridges have been carried out by visual inspection which is time consuming, expensive and unreliable. Now, fiber optic crack sensors are used for crack detection with accuracy.

6.4 MOIRE FRINGES

The term "moiré" is a French word referring to an irregular wavy finish seen on a fabric when pressed between engraved rollers. A beat pattern produced between two gratings of approximately equal spacing is also known as moiré pattern. Moire pattern is seen in daily life on the overlapping of two window screens, the rescreening of a half-tone picture or with a striped shirt seen on television. Photographs of a TV screen taken with a digital camera often exhibit moire patterns.

Figure 6.4 Moire patterns

Figure 6.4 illustrates formation of moire fringes. One grid consists of equally spaced straight lines sloping upwards to the left. The grid sloping upwards to the right consists of straight lines with same spacing as left grid. The pattern is relatively bright around the points of intersection of the two grids. The loci of these points specify the positions of the moire fringes. If the gratings are not identical straight-line gratings, the moiré pattern (bright and dark fringes) will not be straight equi-spaced fringes. Moire patterns are extremely useful to understand basic interferometry and interferometric test results.

More complicated gratings such as circular gratings can also produce Moire fringes. Figure 6.5 shows the superposition of two circular line gratings. This pattern indicates the fringe positions obtained by interfering two spherical wave fronts. The centers of the two circular line gratings can be considered the source locations for two spherical waves.

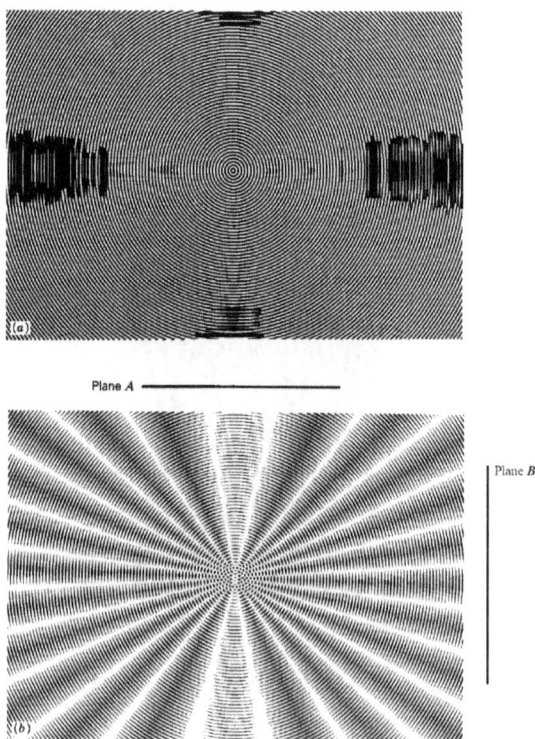

Figure 6.5 (a) Circular line grating representing a spherical wavefront. (b) Moire pattern obtained by superimposing two circular line patterns. (After *Creath. O. K, and Wyant. J., 1992*)

These techniques can be used for displacement measurement or stress analysis as well as for contouring objects. Displacement measurement is performed by comparing the fringe patterns obtained before and after a small movement of the object or before and after applying a load to the object. Time-average vibration analysis can also be performed with moire, yielding results similar to those obtained with time-average holography with a much longer effective wavelength. In manufacturing industries, these patterns are used for studying microscopic strain in materials.

6.5 SENSOR TYPES

In the last two decades, the proliferation of fiber optic sensors and sensing systems continuously increases due to its advantages in contrast to conventional ones. Fiber optic sensors technology offers the possibility of sensing different parameters like displacement, strain, temperature, pressure and flow rate in harsh environment and remote locations. These sensors modulates some features of the light wave in an optical fiber such an intensity and phase or use optical fiber as a medium for transmitting the measurement information. Fiber optic sensors play an important role in improving industrial processes, quality control systems, medical diagnostics and preventing and controlling general process abnormalities. One should not confuse the terms transducers and sensors. A transducer is a device that converts a primary form of energy (mechanical, thermal, electromagnetic, optical, chemical, etc.) into a corresponding signal with a different energy. On the other hand, a sensor is a device that detects and measures a signal or stimulus.

6.5.1 Pressure Sensor

Pressure sensors for engine manifolds, fuel lines, exhaust gases, tires, seats and other uses are common in automotive applications. In biomedical applications, pressure sensors are used in implantable devices for measuring ocular, cranial or bowel pressure and devices built into catheters that can assist in procedures such as angioplasty. Many industrial applications related to monitoring manufacturing processes utilize pressure sensors.

Pressure sensors with optical pick-off are most popular due to their small size and light weight, immunity to electromagnetic interference and harsh environments. Fiber Optic pressure sensors either detect a change in the intensity of the reflected light or detect a change in the phase of the reflected light. The former are classified as intensity based sensors and the latter as interferometric sensors. There are different types of interferometers viz, the Mach-Zehnder interferometer, the Michelson interferometer, the Sagnac interferometer and the Fabry–Perot interferometer (FPI).

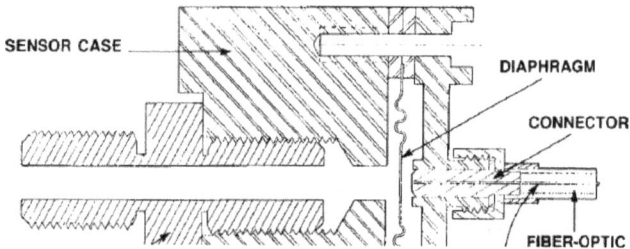

Figure 6.6   A simple fiber optic pressure sensor

A simplest fiber sensor for measuring pressure is shown in Figure 6.6. This sensor takes the advantage of the increased attenuation experienced by the fiber when it bends. A bare length of optical fiber is sandwiched between two serrated (row of sharp points along the edge) pieces of rubber or plastic matting. The fiber is straight and the light detector is at the end of the fiber. If one steps on the mat, bends are created at the sharp point along the edge and hence the light intensity is reduced. This is detected by the detector at the end of the fiber. By changing the size of the serrations and the materials, it is obviously possible to change the sensitivity of the device to detect a wide range of pressures. This sensor is also known as microbend sensors.

Various approaches for measuring pressure with optical fibers have been reported in the literature. These may be broadly grouped as static and dynamic sensors. Static sensors are generally intensity-modulated devices. Interferometric techniques are generally restricted to dynamic measurements (acoustic). This is due primarily to low-frequency noise arising from temperature and $1/f$-type noise. For high-frequency and acoustic measurements, mechanical resonances limit the response.

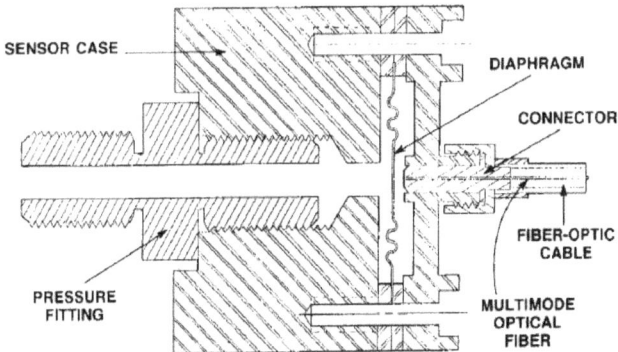

Figure 6.7 Fiber optic pressure sensor design (Courtesy: C.M. Davis and C.J. Zarobila, in Fiber optic sensors, Optoelectronics Materials, McGraw Hill)

Some of the techniques for the measurement of static pressure are transmissive, moving grating, near total internal reflection, reflection and microbending. Static pressure sensors utilizing the reflection technique are conceptually simple and can be grouped into two general classes: one employing a single lead fiber and the other a pair of fibers (or bundle). The intensity versus displacement Calibration curve is required to relate the displacement of the reflector to the pressure. A typical sensor head design is shown in Figure 6.7. The pressure port is at the left and the optical port is at the right. A diaphragm or some other pressure-sensitive transducer (e.g., bourdon tube or bellows) separates the pressure and optical regions. In order to isolate the optical side of the diaphragm from the environment, a permanent plug is used in the optical connector. A section of optical fiber may serve as the plug. Since intensity fluctuations are interpreted as pressure fluctuations, it is important to use a well-regulated optical source. Without such regulation, fluctuations of 10 percent or more may occur, thereby limiting the accuracy of the sensor. Lead perturbations due to microbending or temperature fluctuations can cause errors as large as several percent.

Dynamic pressure sensors have been developed for acoustic applications. Most employ interferometric techniques because of their high sensitivity and large dynamic range. The optical fiber in the sensing arm of the interferometer is caused to undergo a mechanical strain arising from the acoustic pressure. This strain can be significantly increased over that of a bare fiber by the use of either a compliant jacket on the fiber or a compliant mandrel around which the fiber is tightly wound. The resulting mechanical strain causes a phase shift.

Disposable blood-pressure sensor
Figure 6.8 illustrates the typical disposable blood pressure sensor. The disposable blood pressure sensor is made of clear plastic so that the air bubbles can be seen. Saline flows from the intravenous (IV) bag through the clear intravenous tubing and the sensor to the patient. This flushes blood out of the tip of the indwelling catheter to prevent clotting. A lever can open or close the flush valve. The silicon chip has a silicon diaphragm with a four-resistor wheatstone bridge diffused into it. Its electrical connections are protected from the saline by a compliant silicone elastomer gel which also provides electrical isolation. This prevents electric shock from the sensor to the patient and prevents destructive currents during defibrillation from the patient to the silicon chip.

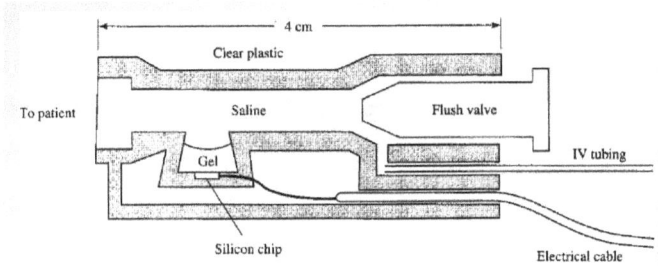

Figure 6.8 Disposable blood pressure sensor

Fiber optic pressure sensor play an important role in pressure measurement in medical devices in life sciences and biopharmaceutical industries since it is intrinsically safe and the optical pressure transducer provide dynamic frequency response time with minimal temperature shift and moisture induced drift. Opsen's miniature and tip sensing fiber optic sensor (0.25 mm and 0.40 mm) is ideal sensor for making cardiovascular blood pressure catheter. The ability in size reduction allows lesser invasive catheterization practices, enables new practices that were not possible with conventional blood pressure measurement system.

6.5.2 Temperature Sensor

A variety of techniques for the measurement of temperature using optical fibers have been reported in the literature. Fiber optic temperature sensors generally refer to those devices measuring higher temperatures. Lower temperature from –100°C to 400°C can be measured by activating various sensing materials such as phosphors, semiconductors or liquid crystals with fiber optic links. Phase-modulated fiber-optic sensors have been shown to be a highly sensitive means for measuring temperature. Fiber optic temperature sensors have proven invaluable in measuring temperatures in basic metals and glass productions as well as in the initial hot forming processes for such materials. Boiler burner flames and tube temperatures as well as critical turbine areas are typical applications in power generation operations.

Rolling lines in steel and other fabricated metal plants also pose harsh conditions which are well handled by fiber optics. Typical applications include furnaces of all sorts, sintering operations, ovens and kilns. Automated welding, brazing and annealing equipment often generate large electrical fields which can disturb conventional sensors. High temperature processing operations in cement, refractory and chemical industries often use fiber optic temperature sensing. At somewhat lesser temperatures, plastics processing, paper making and food processing operations are making more use of the technology. Fiber optics is also used in fusion, sputtering and crystal growth processes in the semiconductor industry.

Beyond direct radiant energy collection or two-color methods, fiber optic glasses can be doped to serve directly as radiation emitters at hot spots so that the fiber optics serves as both the sensor and the media. Westinghouse has developed such an approach for distributed temperature monitoring in nuclear reactors. A similar approach can be used for fire detection around turbines or jet engines. Internal "hot spot" reflecting circuitry has been incorporated to determine the location of the hot area. Fiber optic temperature sensors can be classified as fluorescent intensity-based and interferometric sensors.

Fluorescence decay temperature sensor   In this type of sensor, fluorescent material is used in conjunction with optical fiber. The fluorescent light decay time is used to measure the ambient temperature (Bosselmann et al. 1991). Exciting light is guided through the fiber and from which the fluorescent light is transmitted to a detector. The detector signal is fed back to the modulation of the emitter and thus obtains a self-oscillating system. The frequency of its oscillations depends on the fluorescence decay time and hence the temperature.

The fluorescence decay time varies between 340 μs at 0°C and 132 μs at 100°C. The oscillation frequency can be easily measured with high accuracy and transmitted across long distance without any danger of interference. The accuracy of this method has been reported at 0.25°C for high temperature ranges. Alternatively, instead of monitoring this decay directly using short pulse excitation and accurate timing, the same information can be extracted as a phase delay between a sinusoidally modulated input and the emergent sinusoidally modulated fluorescence.

Blackbody radiation temperature sensor In blackbody radiation temperature sensing practice, the device involves a blackbody cavity formed on the tip of a thin (0.05-0.3 m long) single crystal aluminum oxide (sapphire) fiber. The radiance emitted from the cavity in a narrow wavelength band is used to measure its temperature. The high-temperature sapphire fiber and a low temperature connecting fiber transmit the signal to an interference filter and an optical detector. This type of device can be operated above 600°C with deviations of 0.050% to 0.072%. The blackbody radiation temperature sensor is shown in Figure 6.9.

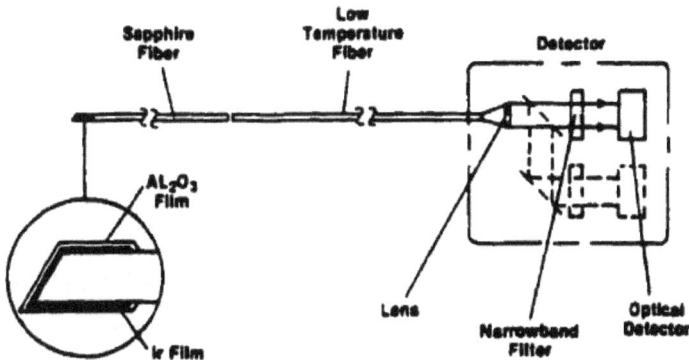

Figure 6.9 Blackbody Radiation Temperature Sensor

Interferometric Temperature Sensors The simplest fiber optic thermometer is the Michelson sensor (Farahi, F.,1993). With a Michelson interferometer, light from a laser is amplitude divided by a fiber coupler to produce reference and signal beams which in turn propagate in the arms of the interferometer.

Figure 6.10 shows the configuration of this interferometer. After traversing the interferometer arms, the beams are recombined at the coupler where coherent mixing takes place with the output being monitored by a photo detector.

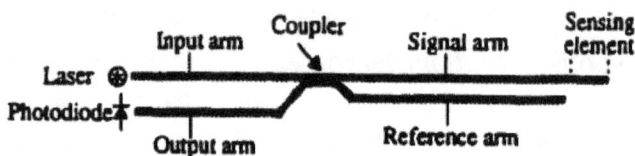

Figure 6.10   Fiber Optic Michelson interferometer sensor.

The principle behind interferometric sensing is that a measurand induced change in the index of refraction n (and hence the propagation constant β) and the length **L** may be recovered by corresponding change in the phase retardation

$$\phi = \frac{2\pi L n}{\lambda} = \beta L \tag{1}$$

The external field may affect either L or n, and the change in ϕ may be described as

$$\frac{I}{L}\frac{\delta\phi}{\delta x} = \frac{\beta}{L}\frac{\delta L}{\delta x} + \frac{\delta\beta}{\delta x} \tag{2}$$

With the available optical detectors, it is not possible to recover the optical phase directly. Hence, the phase may be obtained indirectly by interfering the phase modulated beam with the mutually coherent light of a reference beam.

The intensity of the interference signal between interferometer is given by

$$I = I_0\left[1 + V\cos\varphi\right] \tag{3}$$

where, I_o is the intensity constant and V ≤ 1 is the visibility of the fringes at the output of the interferometer. The equation (3) shows that the measured induced phase change given by equation (2) produces intensity modulation at the output of the interferometer. This is the mechanism in which the effect of external fields are measured interferometricaly, and thus a shift in the temperature causes a change in the signal phase which shifts the intensity of the interference fringes. It is this measured shift that indicates the temperature change.

A miniature extrinsic fiber-optic based Fizeau interferometric temperature sensor with a typical cavity length of several hundred microns has been reported by Rao, Y.J. et al, 1994. The resolution using pseudo-heterodyne phase detection was 0.006°C over a range of 35°C. The temperature probe is separate from the other probes but is integrated by the same local receiving interferometer by exploiting simple spatial multiplexing topologies. This particular temperature probe was assembled by bonding the coated multimode fiber to the inside of a stainless-steel tube which was in turn slipped over the single-mode fiber and bonded when the desired Fizeau cavity width was obtained. Fizeau Interferometric Low-Temperature Sensor is shown in Figure 6.11.

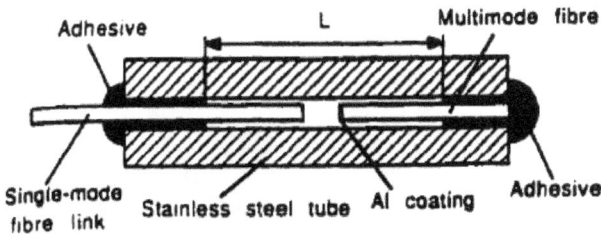

Figure 6.11 Schematic of Fizeau Interferometric Low-Temperature Sensor

An activated temperature measuring system involves a sensing head containing a luminescing phosphor attached at the tip of an optical fiber. A pulsed light source from the instrument package excites the phosphor to luminescence and the decay rate of the luminescence is dependent on the temperature. These methods work well for non-glowing but hot surfaces below 400°C.

Gallium Arsenide based fiber optic temperature sensor offers the highest reliability and robustness for sensing temperature. It uses the well established temperature transduction mechanism technique based on the temperature-dependent bandgap of GaAs crystal. The sensor GaAs crystal encapsulated in uniform protective tubing provides robustness and small size and ensures full protection against mechanical stress. The typical applications of Gallium Arsenide based fiber optic temperature sensor includes

    a. hot spot temperature monitoring of power transformer windings,

    b. temperature monitoring of power transformer top oil and

    c. temperature monitoring at high voltage environments.

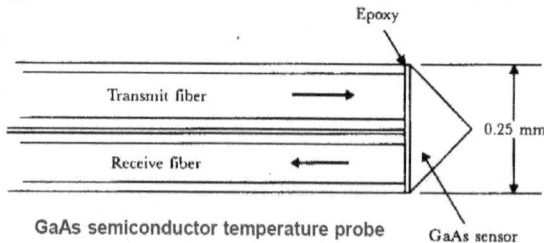

Figure 6.12   GaAs semiconductor temperature probe

Temperature plays an important role in biological applications especially in human physiology. Low body temperature indicates the onset of problems such as stroke and high body temperature indicates infection. Adverse temperature in the body can destroy temperature sensitive enzymes and proteins. In many treatment plans, temperature measurement and regulation is critical. The fiber optic temperature sensor plays a vital role in measuring human temperature precisely. The sensor operation in fiber optic temperature sensor is as follows. A small prism-shaped sample of single crystal undoped GaAs is attached to ends of two optical fibers. The light energy absorbed by the GaAs crystal depends on temperature, and the percentage of received energy versus transmitted energy is a function of temperature. This sensor can be made small enough for biological implantation. Figure 6.12 is the schematic diagram of GaAs semiconductor temperature probe.

Temperature sensors employing Fabry-Perot interferometers whose dimensions are sufficiently small to allow the use of very short coherence length sources (e.g., LEDs) are available from Metricor [Saaski, E., J. Hartl, and G., Mitchell, 1988]. These sensors employ a Fabry–Perot cavity containing a material whose optical refractive index is large and temperature-dependent. Changes in temperature vary the refractive index and the optical phase.

An OPSENS fiber optic temperature sensor uses white light interferometry (WLPI) technology and the sensor is highly robust. It offers a resolution of 0.05°C and an excellent accuracy of ± 0.15°C. The Gallium Arsenide optical sensor is the smallest optical sensor with a dimension of 0.17 mm. It offers a fast response time of less than 10 ms. With an accuracy of ± 0.3°C and resolution of 0.05°C, it is designed to meet the requirements for the life sciences and medical industry. Opsens fiber optic temperature sensors is available as surface sensing probe offering superior surface temperature measurement and most robust packaging for oil filled transformer winding temperature monitoring. Sensors is available as surface sensing probe offering superior surface temperature measurement and most robust packaging for oil filled transformer winding temperature.

6.5.3 STRAIN SENSOR

There are a variety of applications within the automotive, aerospace, electrical power, industrial process as well as the oil and gas industries that call for strain sensors capable of measuring static and dynamic stresses under harsh environments and at elevated temperatures in the 200°C – 800°C range. Some examples of typical applications are strain monitoring. At these elevated temperatures, conventional foil strain gauges cannot operate. Optical fiber sensors offer the possibility to perform strain and temperature measurements under harsh conditions in the presence of electrical noise, EMI interference and mechanical vibrations. Intrinsically, silica-based glass fibers are capable of operating up to 800°C after removing their protective polymeric coatings. For higher temperatures, sapphire fibers can be used, allowing their operation at temperatures as high as 1500°C.

The strain sensors available possess outstanding repeatability, temperature independent, insensitive to transverse strains, electromagnetic and radiofrequency interference and microwave immune, intrinsically safe and miniature in size. They can be applied in EM, RF and microwave environments, high voltage environments, nuclear and hazardous environments and civil engineering and geotechnical applications.

Opsens designed a miniature and robust fiber optic strain gauge sensor which uses two optical fibers that are precisely aligned inside a micro capillary tube to form an optical Fabry-Pérot interferometer. This makes this strain gauge completely immune to any electromagnetic interference. It is also completely insensitive to transverse strains and temperature, as opposed to fiber Bragg gratings. It can be used in the temperature range from −40°C to +250 °C.

Recently, Tian Zhao et al. (2011) reported microextrinsic fiber-optic Fabry-Perot interferometric (MEFPI) strain sensor. The experimental setup used for testing the strain responses of the MEFPI sensor is shown in Figure 6.13. The reflection of the MEFPI sensor was monitored by using a high-accuracy optical spectrum analyzer with a wavelength scanning range of 1510–1590 nm and a wavelength resolution of 2.5 pm. They measured the strain responses of the graded-index multimode fiber (GI-MMF)-based MEFPI sensors at room temperature (25°C).

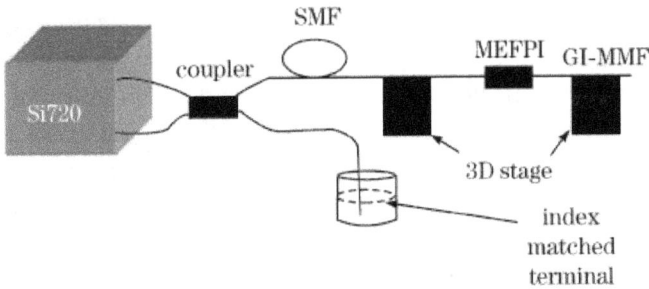

Figure 6.13 Schematic diagram of the experimental setup for measuring strain response of MEFPI sensor

6.5.4 LIQUID LEVEL SENSOR

Industry frequently needs to measure liquid levels in tanks and other large containers. This need arises not only in heavy industry where great volumes of liquids are often stored but also in light industry, such as olive-oil pressing. Further, the large-scale operations such as public water supplies and treatment plants, fuel storage for public-transport systems and service stations also require liquid level sensors. These liquids may be inert (as in the case of water) or highly flammable (as in the case of many petroleum derivatives). In the case of gasoline stations, a common measurement method is to plunge a measuring rod into the underground tank to determine the level of fuel. This rudimentary method tends to be slow and inefficient. In automobiles or ships, the fuel is measured by a float connected to a variable resistance, indicating the level of the liquid inside the tank. The main disadvantage of this system is that an electrical current (though weak) must be introduced into the flammable or simply conducting liquid.

The Fiber-optic liquid-level continuous gauge system designed by F. Perez-Ocon et al. 2006, has a great advantage over other methods because it is non-electrical and is also immune to electromagnetic interference. This is an intrinsic and intrusive measuring device for precise continual monitoring of liquid levels. The optical fiber is the only part of the measuring device introduced into the tank. That is, only light and plastic or glass interacts with the liquid and hence the method becomes safe and corrosion free, without electrical sparks that could cause a fire or explosion of a tank. In addition, the measurement is highly sensitive, accurate and repeatable. There are no moveable parts (no wear and no problem with replacement pieces), no problems of hysteresis (repeatable) and the light is always guided in the same way within the fiber (repeatable and accurate). Moreover, for the characteristics of any fiber optics, any variation, no matter how small, in the cladding, is manifested as a great variation in the light that reaches the end face. The general scheme of the gauge to measure the levels of liquid is shown in Figure 6. 14.

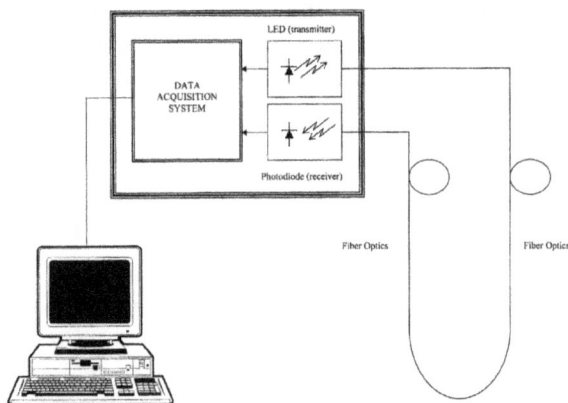

Figure 6.14 General scheme of the gauge to measure the levels of liquid

The light from a LED is directed into an optical fiber immersed in the liquid whose level one wants to measure. The light travels through the fiber. When the fiber is completely surrounded by liquid, the liquid acts as a second cladding. Thus the total internal reflection occurs at the core–cladding interface, and the light reaches the receptor practically without loss except for absorption. When the tank is not completely full of liquid, one part of the fiber is surrounded by liquid and the other part is surrounded by air. In the submerged portion, total internal reflection occurs because the fiber has a cladding (cladding of the fiber plus the liquid) and the losses of light are very small because most of the rays are bounded. In the exposed portion of the fiber less total internal reflection occurs because the fiber has only the cladding and there are more leaky rays. If we calibrate the gauge for the different liquid heights in the container, then simply by reading the outgoing signal of the detector we can determine the height of the liquid. The higher the level, the larger value in the reading of the photodiode. Here, the sensor is composed of four main parts, namely, the light source (LED), the fiber, the photodiode (PIN) and the data-acquisition system (DAS). The data-acquisition system (DAS) consists of a signal conditioner, a digital/analogical converter (DAC), a microcontroller (PIC) and a computer (PC).

Fiber optic sensing can detect and localize leaks continuously and accurately all along the length of the pipeline. It does this by detecting the change of temperature which occurs when a pipe leaks.

There are two types of tanks, viz, 'on the ground tank' and under ground tank for storing gasoline. If there is any leakage 'on the ground tank', it can be visually seen and rectified. If it is underground, any leakage can spoil the environment. Fiber optic sensors are widely used to detect such leakages by winding fibers on the tank surface at the time of designing the tank.

6.6 SENSORS IN MEDICAL FIELD

6.6.1 Endoscope

Endoscopes are used to look inside the body where one cannot reach. An endoscope consists around 50,000 very thin fibers of 8 μm diameter, each carrying a single light level. An endoscope is about one meter in length with a diameter of about 6 mm or less. For illumination, some of the fibers are used to carry light from a 300 watt xenon bulb. A lens is used at the end of the other fibers to collect the picture information which is then displayed on a video monitor for easy viewing. To rebuild the image at the receiving end, it is essential that the individual fibers should maintain their relative positions within the endoscope. Other wise the light information will become distorted scrambled. Bundle of fibers in which the position of each fiber is carefully controlled are called coherent bundles.

6.6.2 Medical Applications

Fiber optics has been used in the medical applications for years. The physical characteristics of fiber make it a natural choice for many different applications. Commonly used for illumination, flexible image bundles, light conductors, flexible light guides, laser delivery systems and equipment interconnects. Fiber optics provides a very compact, flexible conduit for light or data delivery in equipment, surgical and instrumentation applications. Optical fibers are reliable, safe, easy to handle, sterilizable, high performance, insensitive to magnetic field and light weight.

Traditional medical fiber optic applications include light therapy, x-ray imaging, ophthalmic lasers, laboratory and clinical diagnostics, dental hand pieces, surgical and diagnostic instrumentation, endoscopy, surgical microscopy and a wide range of equipment and instrument illumination.

Medical instruments utilize fiber optics for a variety of applications including illumination, image transfer, and laser signal delivery. A large portion of the fiber used in these applications support site illumination either as an integrated component of an instrument or as an individual light source such as examination lights, headlight, laryngoscope (blade illumination), anoscope (with annular illumination), otoscope, binocular indirect ophthalmoscope, amnioscope, microscope illumination and heart catheter.

Fiber optic technology is opening exciting new areas of medical fields day by day. The physical characteristics of fiber make it a natural choice for many uses. Optical fiber has become widely used in imaging, laser delivery systems, illumination, sensors and equipment interconnects. Optical fibers provide a compact and flexible conduit for light or data delivery in diagnostic and interventional medical applications.

Fiber-optic biomedical sensors (Courtesy: Laser focus world) Optical fiber sensors comprise a light source, optical fiber, external transducer and photo detector. They sense by detecting the modulation of one or more of the properties of light (intensity, wavelength, or polarization) that is guided inside the fiber. The modulation is produced in a direct and repeatable fashion by an external perturbation caused by the physical parameter to be measured. The measurand of interest is inferred from changes detected in the light property. Biomedical FOS can be categorized into four main types

i.   physical,

ii. imaging,

iii. chemical and

iv. biological.

Physical sensors measure a variety of physiological parameters like body temperature, blood pressure and muscle displacement. Imaging sensors encompass both endoscopic devices for internal observation and imaging as well as more advanced techniques such as optical coherence tomography (OCT) and photo-acoustic imaging where internal scans and visualization can be made nonintrusively. Chemical sensors rely on fluorescence, spectroscopic and indicator techniques to identify and measure the presence of particular chemical compounds and metabolic variables (such as pH, blood oxygen, or glucose level). They detect specific chemical species for diagnostic purposes as well as monitor the body's chemical reactions and activity. Biological sensors tend to be more complex and rely on biologic recognition reactions such as enzyme-substrate, antigen-antibody, or ligand-receptor to identify and quantify specific biochemical molecules of interest.

In terms of sensor development, the basic imaging sensors are the most developed. Fiber optic sensors for measurement of physical parameters are the next most prevalent and the least developed area in terms of successful products is sensors for biochemical sensing, even though many FOS concepts have been demonstrated. Applications for biomedical sensors can be classified as *in vivo* or *in vitro*. *In vivo* refers to application on a whole, living organism such as a human patient. *In vitro* refers to testing outside the body such as laboratory blood tests. From the perspective of how sensors are applied to a patient or biological system, they can be classified as noninvasive, contacting (skin surface), minimally invasive (indwelling) or invasive (implantable). Biomedical sensors can be used in humans (clinical), in animals (veterinary) or other living organisms (life sciences). It can be used for diagnostic, therapeutic or intensive care in clinical applications. The classification of biomedical sensors by type showing various biomedical parameters of interest is given in Table 6.1.

Table 6.1 The classification of biomedical sensors by type showing various biomedical parameters of interest

Physical	Chemical	Biological	Imaging
Body temperature	pH	antigens	Endoscopy
Blood pressure	pO_2	antibodies	Optical coherence tomography (OCT)
Blood flow	PCO_2	electrolytes	Photodynamic therapy
Heart rate	Oximetry (SaO_2, SvO_2)	Enzymes	
Force	Glucose	Inhibitors	
Position	Bile	Metabolities	
Respiration	Lipids	Proteins	

Sensors intended for implanting or indwelling applications must be very small (as shown in Figure 6.15) such as this micro-miniature fiber-optic pressure sensor shown on a fingertip.

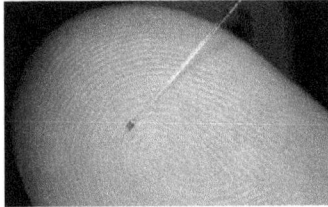

Figure 6.15 Sensors intended for implanting or indwelling applications must be very small such as this micro-miniature fiber-optic pressure sensor shown on a fingertip *(Courtesy of Samba Sensors AB)*

Fiber Bragg grating sensors are mounted on the tip of an intra-aortic catheter that also serves as a laser-ablation delivery probe for the treatment of atrial fibrillation. The FBGs detect the force exerted against the heart wall by the stress induced on them (Figure 6.16). Force control is essential for delivering appropriate laser ablation pulses needed to produce lesions that are induced in the heart walls to reduce abnormal electric activity.

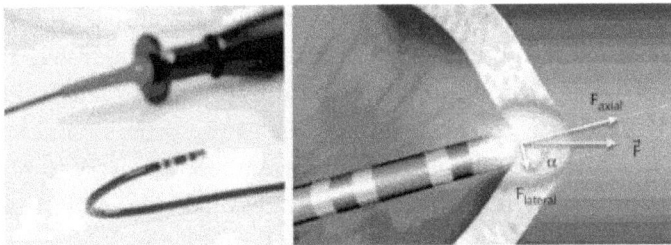

Figure 6.16 A fiber-optic intra-aortic force sensing catheter probe enables real-time monitoring of the force exerted against the heart wall by the catheter *(Courtesy of EndoSense)*

Combing FOS with lasers proved beneficial and successful in treating kidney and urethra stones, removal of short circuit in palpitation, neuro surgery and ablation of tumors etc.,

Disposable sensors in medical field Disposable OFSs, however, may represent a unique opportunity for this emerging technology. Recent advances in minimally invasive medical technologies, such as those that employ disposable instrumented catheters, demand smaller and more reliable sensors. These sensors should be insensitive to EM interference from surrounding equipment and surgical tools that use microwave or RF probes. This is especially important now that magnetic resonance imaging (MRI) of soft tissue is coming into wider use in operating theaters. Integrating miniature OFSs in state-of-the-art catheters can meet these challenges. Still, integrating OFS technology into disposable medical tools is not trivial. It involves a variety of interdependent issues including optical technology selection, biocompatible, unaffected by the particular sterilization methods, resistant to damage from storage, product design, packaging materials, cost-effective manufacturing and quality control.

6.7 NEW FIBERS

The introduction of lasers into medicine has made many surgeries less invasive and more accurate. The procedures that could not be performed earlier can be done now. The fiber overcomes medical challenges. Lasers require a delivery system to transport the light to the surgical site. Hence, bulky articulated arms with refelecting mirrors were used for the purpose. Now optical fiber replaced the above system with portable and easy-to-use probes (similar to surgical cutting tools). Fibers have even allowed lasers to reach areas previously unreachable with traditional surgery. But laser surgery poses a great many challenges for fiber optics. Fiber must be safe for both surgeon and the patient. Secondly, it should withstand sterilization, high power laser (in case of surgery) and bending. Thirdly, there are even more challenges for each specific operation due to the new laser sources and hence new fibers.

The desirable wavelength for silica fibers is 1.3 to 1.6 μm At 1.4 μm OH absorption is dominated and hence, one has to avoid

1.4 µm. At 1.6 µm, silica fiber completely absorbs the radiation. Hence, it is not suitable for the λ = 1.6 µm. Hence, we require new materials for optical fibers. Further, long wavelength lasers are available at present. They are made from materials such as germanium-oxide glass, heavy metal fluoride glass, sapphire, polycrystalline and hollow waveguide fibers. An important feature of current IR fibers transmits wavelengths longer than most oxide glass fibers. Although they have the capability of transmitting beyond 20 µm, most applications do not require the delivery of radiation longer than 12 µm. While none of these fibers had physical properties even approaching that of conventional silica fibers, they were nevertheless, useful in lengths less than 2 to 3 m for a variety of medical applications, IR sensors and power delivery applications. IR fibers are much weaker than silica fibers and therefore more fragile. Mid IR lasers such as Er:YAG (2.94 µm) and Er:YSGG (2.79 µm) has been developed based on germanium – oxide glasses.

GeO_2 – based glass fibers are composed of GeO_2 (30-76%) - RO (15-43%)-XO (3-20%) where R- represents an alkaline earth metal and X represents an element of group IIIA.

The special features of GeO_2 fibers are:

1. Germanium oxide fiber is stronger & reliable

2. They can handle more power with excellent durability

3. They are chemically durable than fluoride fibers

4. Optical cladding minimizes bending losses. Fiber transmission losses are very low- less than 0.7 dB/m at 2.94 µm and 0.25 dB/m at 2.79 µm with power handling over 20W for a 450-µm core fiber.

5. Possesses high refractive index of 1.84 (bending losses are very high in sapphire fiber and hollow wave guide)

IR fibers can transmit wavelengths longer than most oxide glass fibers. IR fibers are much weaker than silica fiber and, therefore, more fragile. Therefore, their use is restricted to applications in chemical sensing, thermometry, and laser power delivery. The categories of IR fibers and their applications are listed in Table 6.2 and 6.3 respectively.

Table 6.2 Categories of IR fibers

Material	Fiber	Example
Glass	Heavy metal fluoride - HMFG	ZBLAN - ($ZrFM_4$-BaF_2-LaF_3-AlF_3-NaF)
	Germanate	GeO_2-PbO
	Chalcogenide	As_2S_3 and AsGeTeSe
Crystal	Polycrystalline - PC	AgBrCl
	Single crystal - SC	Sapphire
Hollow waveguide	Metal/dielectric film	Hollow glass waveguide
	Refractive index < 1	Hollow sapphire at 10.6 µm

Table 6.3 Applications of IR fibers

Application	Suitable IR fibers
Fiber optic sensors	AgBrCl, sapphire, chalcogenide, HMFG
Fiber optic chemical sensors	Hollow glass waveguides
Radiometry	Hollow glass waveguides, AgBrCl, chalcogenide, sapphire
Er:YAG laser power delivery	Hollow glass waveguides, sapphire, germanate glass
CO_2 laser power delivery	Hollow glass waveguides
Thermal imaging	HMFG, chalcogenide
Fiber amplifiers and lasers	HMFG, chalcogenide

Scientists have created a new type of fiber optic cable with a zinc selenide core, which is said to be better than conventional cables (Figure 6.17) at transmitting and manipulating light. A team of scientists led by John Badding, a professor of chemistry at Penn State University, has developed the very first optical fiber made with a core of zinc selenide - a light-yellow compound that can be used as a semiconductor. To make the fibers, the scientists started with hollow glass capillaries. Using a unique high-pressure chemical-deposition technique, they were then able to deposit the zinc selenide waveguiding cores inside of them. The new class of optical fiber, which allows for a more effective and liberal manipulation of light, promises to open the door to the development of improved surgical and medical lasers. Zinc selenide optical fibers also may open new avenues of research that could improve laser-assisted surgical techniques, such as corrective eye surgery. Zinc selenide uses nonlinear frequency conversion and hence is more efficient at converting light from one color to another. Since Zinc selenide fiber can transmit wavelengths of up to 15 microns, it will be useful for laser-radar technology. The detection of pollutants and environmental toxins could be yet another application of better laser-radar technology capable of interacting with light of longer wavelengths.

Figure 6.17 Scientists have created a new type of fiber optic cable with a zinc selenide core, which is said to be better than conventional cables at transmitting and manipulating light (Photo: Beria)

The Next Generation of Optical Fibers is photonic fibers. MIT's new optical fiber (photonic fibers) carries more power with less loss. The new fiber would allow a carbon dioxide laser's high power to be transmitted over longer distances than are possible today. The possible applications include medical treatments that necessitate high-power delivery such as surgery or facilitating the breakup of kidney stones and medical diagnosis requiring broad-band infrared transmission such as detecting cancerous cells. In the manufacturing and materials processing, the fiber may play a crucial role.

6.7.1 Future Trends for Sensor technology

Special waveguides such as photonic crystal fibers will enable many new sensing mechanisms and sensor configurations. Improved micro-fabrication technologies will continue to improve sensor performance, functionality, reliability and capability of harsh environment operation. Advanced signal processing and network technology will enable high-density fiber optic sensor networks.

REVIEW QUESTIONS

1. Classify the types fiber optic sensors

2. List he advantages of fiber optic sensors

3. With necessary diagram, describe a fiber optic sensor system.

4. Describe in detail the applications of fiber optic sensors.

5. What are Moire fringes? How it can be obtained? Mention their applications.

6. Describe a simple fiber optic pressure sensor.

7. Write a note on disposable blood pressure sensor.

8. Explain the various types of temperature sensors and describe the salient features of any one of them.

9. With necessary diagram, describe and explain GaAs semiconductor temperature sensor.

10. Draw a schematic diagram of the experimental set up for measuring strain response of MEFPI sensor.

11. What is the importance of liquid level measurement?

12. Draw a block diagram to illustrate the measuring of liquid level.

13. Write a note on the sensors in medical field.

14. What is the importance of new fibers? List the new fibers and its uses.

GLOSSARY

Absorption loss Loss of light in a fiber due to impurities.

Absorption (electromagnetic radiation) Transformation of radiant energy to a different form of energy by the interaction of matter, depending on temperature and wavelength.

AC Abbreviation for alternating current. An electric current that reverses its direction at regularly recurring intervals.

Acceptance angle The largest angle of incident light that lies within the cone of acceptance.

Active medium Collection of atoms or molecules capable of undergoing stimulated emission at a given wavelength.

Adapter A device to join and align two connectors.

Add/Drop The process where a part of information carried in a transmission system is demodulated (dropped) at an intermediate point and different information is entered (added) for subsequent transmission. The remaining traffic passes straight through the multiplexer without additional processing.

Alexandrite A rare oxide mineral and is a color changing variety of the mineral chrysoberyl, $BeAl_2O_4$.

Amplification An increase in power level measured at two points. Usually measures in decibels.

Attenuation The decrease in energy (or power) as a beam passes through an absorbing or scattering medium. Reduction of signal magnitude or signal loss, usually expressed in decibels.

Attenuator 1) In electrical systems, a usually passive network for reducing the amplitude of a signal without appreciably distorting the waveform. 2) In optical systems, a passive device for reducing the amplitude of a signal without appreciably distorting the waveform.

Avalanche diode A device used to convert light into an electrical current.

Bandwidth The range of modulation frequencies that can be transmitted on a system while maintaining an output power of at least half of the maximum response.

BER Bit error rate. The proportion of incoming bits of data that are received incorrectly.

Bidirectional Operating in both the directions.

Bit Abbreviated version of binary digit. The smallest unit of information upon which digital communications are based; also an electrical or optical pulse that carries this information.

Bits per second (bps) The number of bits passing a point every second, the transmission rate for a digital information.

Broadband Service requiring 50–600 Mbps transport capacity.

Byte interleaved Bytes from each STS-1 are replaced in sequence in a multiplexed or concatenated STS–N signal, For example, STS-3 the sequence of bytes from contributing STS1s is 1,2,3, 1,2,3, etc.,

Cavity The laser resonator, or tube, in which the lasing process occurs.

C-Band The wavelength range between 1530 nm and 1562 nm used in some CWDM and DWDM applications.

CDMA Abbreviation for code-division multiple access. A coding scheme in which multiple channels are independently coded for transmission over a single wideband channel using an individual modulation scheme for each channel.

Channel The smallest subdivision of a circuit that provides a type of communication service, usually a path with only one direction.

Channel capacity Maximum number of channels that a cable system can carry simultaneously.

Chromatic dispersion Dispersion caused by different wavelengths contained in the transmitted light traveling at different speeds.

Chromium-doped gain media Laser gain media doped with chromium ions.

Circuit A communication path or network, usually a pair of channels providing bidirectional communication.

Circuit switching Basic switching process whereby a circuit between two users is opened on demand and maintained for their exclusive use for the duration of the transmission.

Cladding The clear material surrounding the core of an optic fiber. It has lower refractive index than the core.

Coarse Wavelength–division Multiplexing (CWDM) – CWDM allows eight or fewer channels to be stacked in the 1550 nm region of optical fiber, the C-Band.

Coherent bundles A group of optic fibers in which each fiber maintains its position relative to other fibers so that images can be transmitted along the bundle.

Coherent light Light waves that are "in phase" with one another. Monochromaticity and low divergence are two properties of coherent light.

Continuous Wave (CW) Constant, steady-state delivery of laser power. The continuous-emission mode of a laser as opposed to pulsed operation.

Core The central part of the fiber through which most of the light is transmitted. It has a higher refractive index than the cladding.

Coupler A device to combine several incoming signals onto a single fiber or to split a single signal onto several fibers in a predetermined power ratio.

Critical angle The angle of incidence above which total internal reflection occurs.

Crystal A solid with a regular array of atoms. Sapphire (Ruby Laser) and YAG (Nd:YAG laser) are two crystalline materials used as laser sources.

dB (decibel) A logarithmic unit used to compare two power levels.

dBm Abbreviation for decibel relative to milliwatt.

Defect A limited interruption in the ability of an item to perform a required function.

Demultiplexer A module that separates two or more signals previously combined by compatible multiplexing equipment.

Demultiplexing A process applied to a multiplex signal for recovering signals combined within it and for restoring the distinct individual channels of the signal.

Dense Wavelength–division Multiplexing (DWDM) – The transmission of many of closely spaced wavelengths in the 1550 nm region over a single optical fiber. Wavelength spacings are usually 100 GHz or 200 GHz which corresponds to 0.8 nm or 1.6 nm. DWDM bands include the C-Band, the S-Band, and the L-Band.

Diode An electronic device that conducts a current in only one direction.

Diode Laser A laser that emits coherent light through the injection of electric current into a semiconductor diode.

Diplexer A device that combines two or more types of signals into a single output. Usually incorporates a multiplexer at the transmit end and a demultiplexer at the receiver end.

Directional Lasers emit light that is highly directional, that is; laser light is emitted as a relatively narrow beam in a specific direction.

Dispersion The widening of light pulses on an optic fiber due to different propagation velocities of the pulse components.

Divergence The increase in the diameter of the laser beam with distance from the exit aperture. The value gives the full angle at the point where the laser radiant exposure or irradiance is $1/e$ or $1/e^2$ of the maximum value, depending upon which criteria are used.

Electrical pumping Can be achieved by keeping the laser medium in the electron beam so that the electrons create a population inversion by transferring the energy to the atoms or molecules under collision.

Electromagnetic spectrum The range of all possible frequencies of electromagnetic radiation.

Erbium-doped gain media Laser gain media doped with erbium ions.

Erbium (Er) The fluorescence properties of erbium in various host materials and the stimulated emission around 1.6 μm is of interest to ophthalmologists because the eye is subjected to less retinal damage by laser radiation.

Erbium Laser Cosmetic laser treatment used to reduce the appearance of bags and dark circles under the eyes. Unlike the stronger Fraxel laser, the Erbium Laser only penetrates the first few layers of delicate skin to increase collagen production and reduce the dark circles.

Ethernet The most common Local Area Network protocol.

Excimer (Excited Dimer) A gas mixture used as the active medium in a family of lasers emitting ultraviolet light.

Excitation Energizing a material into a state of population inversion.

Excitation mechanism The excitation mechanism of a laser is the source of energy used to excite the lasing medium. Excitation mechanisms typically used are electricity from a power supply, a flashtube, lamp, or the energy from another laser.

Excited state Atom with an electron in a higher energy level than it normally occupies.

Fabry Perot Generally refers to any device, such as a type of laser diode that uses mirrors in an internal cavity to produce multiple reflections.

Failure A termination of the ability of an item to perform required function, a failure is caused by the persistence of a defect.

Feedback mechanism A laser's feedback mechanism is used to reflect light from the lasing medium back into itself and typically consists of two mirrors at each end of the lasing medium.

Femtoseconds 10^{-15} seconds.

Ferrule A rigid tube used to confine and support the stripped end of a fiber as found in connectors.

Fiber optic cable Flexible glass or plastic strands made into a cable, used to carry light from one place to another.

Flashlamp A tube typically filled with krypton or xenon. Produces a high intensity white light in short duration pulses.

Frequency The number of cycles of periodic activity that occur in a discrete amount of time.

Frequency doubling The phenomenon that an input wave in a nonlinear material can generate a wave with twice the optical frequency.

Frequency-Division Multiplexing (FDM) A method of deriving two or more simultaneous, continuous channels from a transmission medium by assigning separate portions of the available frequency spectrum to each of the individual channels.

Fusion splices A low loss, permanent means of connecting two fibers involving heating the fibers until they fuse together.

GaAlAs laser The most common DPSS laser in use is the 532 nm wavelength green laser pointer. A powerful (>200 mW) 808 nm wavelength infrared

GaAlAs laser diode pumps a neodymium – doped yttrium aluminium garnet (Nd:YAG).

Gain Another term for amplification. A measure of the strength of optical amplification.

Graded index fiber A fiber in which the refractive index of the core is at a maximum value at the center and decrease towards the cladding.

Holmium YAG (Ho:YAG) laser A solid state diode pumped laser with 2.1 μm wavelength useful for tissue ablation, kidney stone removal and dentistry.

I n f r a r e d radiation Electromagnetic radiation of wavelength from 700 nm to 1 mm.

Insertion loss The loss of power due to the insertion of a device, e.g., connector.

Interleave The ability of SONNET to mix together and transfer different types of input signals in an efficient manner, thus allowing higher transmission rate.

Intermodal dispersion The phenomenon that the group velocity of light propagating in a multimode fiber depends not only on the optical frequency (→ chromatic dispersion) but also on the propagation mode involved.

Intramodal dispersion In fiber-optic communication, it is a category of dispersion that occurs within a single-mode.

Jitter Short waveform variations caused by vibrations, voltage fluctuations, control system instability etc.

KTP (Potassium Titanyl Phosphate) A crystal used to change the wavelength of an Nd:YAG laser from 1060 nm (infrared) to 532 nm (green).

LAN (Local area network) A communication network connecting computers and other instruments within a limited area, such as building or groups of buildings like a campus.

LASER An acronym for light amplification by stimulated emission of radiation. A laser is a cavity, with mirrors at the ends, filled with material such as crystal, glass, liquid, gas or dye. A device which produces an intense beam of light of low spectral width with the unique properties of coherence, collimation and monochromaticity.

Laser crystals Crystals used as solid state lasers active media.

Laser device Either a laser or a laser system.

Laser medium (Active Medium) Material used to emit the laser light and for which the laser is named.

Laser pumping The act of energy transfer from an external source into the gain medium of a laser. The energy is absorbed in the medium, producing excited states in its atoms. When the number of particles in one excited state exceeds the number of particles in the ground state or a less-excited state,population inversion is achieved.

Laser resonators Optical resonators serving as basic building blocks of lasers.

Laser rod A solid state, rod shaped lasing medium in which ion excitation is caused by a source of intense light, such as a flash lamp. Various materials are used for the rod, the earliest of which was synthetic ruby crystal.

Laser system An assembly of electrical, mechanical, and optical components which includes a laser.

L-Band L band is also used in optical communications to refer to the wavelength range1565 nm to 1625 nm.

LED (Light emitting diode) A semiconductor device used as a low power light source. The spectral width is greater than laser.

Macrobend A relatively large-radius bend in an optical fiber, such as might be found in a splice organizer tray or a fiber-optic cable that has been bent.

MASER Microwave Amplification by Stimulated Emission of Radiation.

Mechanical splice A permanent method of connecting the fibers usually involving adhesives and mechanical support and alignment of the fibers.

Microbend Microbends are small microscopic bends of the fiber axis. In optical fiber, sharp but microscopic curvatures that create local axial displacements of a few microns and spatial wavelength displacements of a few millimeters.

Mie scattering Mie scattering (after Gustav Mie) that encompasses the general spherical scattering solution (absorbing or non-absorbing) without a particular bound on particle size.

Mode A term used to describe how the power of a laser beam is geometrically distributed across the cross section of the beam. Also used to describe the operating mode of a laser such as continuous or pulsed.

Mode locked A method of producing laser pulses in which short pulses (approximately

10^{-12} second) are produced and emitted in bursts or a continuous train.

Mode-locked lasers Lasers which emit ultra-short pulses on the basis of the technique of mode locking.

Modes Separate optical waves capable of being transmitted along a fiber. The number of modes with a given light wavelength is determined by the NA and the core diameter.

Modulation The ability to superimpose an external signal on the output beam of the laser as a control. A process whereby a feature of a wave, such as amplitude or frequency is changed in order to convey information.

Monochromatic The light emitted from a laser is monochromatic, that is, it is of one color/wavelength.

Monomode fiber Alternative name for single mode fiber.

Multimode fiber An optic fiber able to propagate more than one mode at the same time.

Multiplexer A device for combining several channels to be carried by one line or fiber.

Multiplexing The transmission of several different signals along a single fiber.

Nanometer (nm) A unit of length in the International System of Units (SI) equal to one billionth of a meter. One nm equals 10^{-9} meter, and is the usual measure of light wavelengths. Visible light ranges from about 400 nm in the purple to about 760 nm in the deep red.

Nanosecond One billionth (10^{-9}) of a second. Longer than a picosecond or femtosecond, but shorter than a microsecond. Associated with Q-switched lasers.

Narrowband Service requiring up to 1.5 Mbps transport capacity.

Nd,Cr:GSGG Laser rods made of gadolinium scandium gallium garnet (GSGG) doped with neodymium and chromium, which is important for laser applications in high radiation environments.

Nd:Glass Laser A solid state laser of neodymium: glass offering high power in short pulses. A Nd doped glass rod used as a laser medium to produce 1064 nm light.

Nd:YAG Laser Neodymium Yttrium Aluminum Garnet A synthetic crystal used as a laser medium to produce 1064 nm light.

Neodymium Doped yttrium orthovanadate (Nd:YVO$_4$) – Crystal which produces 1064 nm wavelength light from the main spectral transition of neodymium ion.

Neodymium (Nd) Nd^{3+} was the first trivalent rare earth ions to be used in a laser. It is the active element in Nd:YAG lasers and Nd:Glass lasers.

Network A group of devices such as computers that can communicate with each other.

Node A point where a computer, printer, etc, is connected to a network.

Numerical aperture The sine of the critical angle between the core and the cladding.

Optic fiber The length of clear material that can be used to transmit light. Often abbreviated as fiber.

Optical cavity (Resonator) Space between the laser mirrors where lasing action occurs.

Optical density A logarithmic expression for the attenuation produced by an attenuating medium, such as an eye protection filter.

Optical detectors The components that convert the light energy of fiber optic communications into electrical signals (voltage or current) for recovery of data.

Optical pumping The excitation of the lasing medium (to raise or "pump") electrons from a lower energy level in an atom or molecule to a higher one) by the application of light rather than electrical discharge.

Optically pumped lasers A type of laser that derives energy from another light source such as a xenon or krypton flashlamp or other laser source.

Overhead Extra bits in a digital stream used to carry information besides traffic signals, orderwire, for example, would be considered overhead information.

Packet switching An efficient method for breaking down and handling high volume traffic in a network; a transmission technique that segments and routes information into discrete units; packet switching allows for a efficient sharing of network resources as packets from different sources can all be sent over the same channel in the same bit stream.

Phase Waves are in phase with each other when all the troughs and peaks coincide and are "locked" together. The result is a reinforced wave in increased amplitude (brightness).

Photodiode A semiconductor device that converts light into an electrical current.

Photonics The science and technology of light. It is nothing but the generation and harnessing laser radiations.

Pico A prefix indicating a millionth of a millionth.

Picoseconds A period of time equal to 10^{-12} seconds.

PIN diodes A PIN diode is a diode with a wide, lightly doped 'near' intrinsic semiconductor region between a p-type semiconductor and an n-type semiconductor region. The p-type and n-type regions are typically heavily doped because they are used for ohmic contacts.

Pockel's cell An electro-optical crystal used as a Q-switch for building modulators.

Pointer A part of the SONNET overhead that locates a floating overload structure; STS pointer locate the SPE; VT pointers locate the floating mode VTs; all SONNET frames use STS pointers; only floating mode VTs use VT pointers.

Polarizability The measure of the change in a molecule's electron distribution in response to an applied electric field.

Population inversion A state in which a substance has been energised, or excited, so that more atoms or molecules are in a higher excited state than in a lower resting state. This is a necessary pre-requisite for laser action.

Power budget The maximum possible loss that can occur on a fiber optic system before the communication link fails.

Pulse A discontinuous burst of laser, light or energy, as opposed to a continuous beam. A true pulse achieves higher peak powers than that attainable in a CW output.

Pulse repetition rate The number of pulses emitted per second e.g. by a mode-locked or Q-switched laser.

Pulsed laser Laser which delivers energy in the form of a single or train of pulses rather than continuously.

Pump To excite the lasing medium. (See Optical Pumping or Pumping)

Pumping Addition of energy (thermal, electrical or optical) into the atomic population of the laser medium, necessary to produce a state of population inversion.

Q-factor A measure of the damping of resonator modes.

Q-switch Optical switches that has the effect of a shutter to control the laser resonator's ability to oscillate. These are typically used for generating nanosecond pulses in lasers.

Q-switched laser A laser which stores energy in the laser media to produce extremely short, extremely high intensity bursts of energy.

Q-switching A method for obtaining energetic pulses from lasers by modulating the intracavity losses.

Rayleigh scattering The scattering of light due to small inhomogeneous regions within the core.

Reflection The change in direction of a wavefront at an interface between two different media so that the wavefront returns into the medium from which it originated.

Refraction The bending of light path due to change in refractive index.

Refractive index The ratio of the speed of light in a material compared to its speed in free space.

Regenerator Device that restores a degraded digital signal for continued transmission; also called repeater.

Regenerator Placed at intervals along a digital transmission route, it reconstructs the digital pulse.

Repeater A transmitter and a receiver used at intervals along a transmission route to increase the power in an attenuated signal.

Repetition rate Number of pulses per second produced by a pulsed laser.

Resonator The mirrors (or reflectors) making up the laser cavity including the laser rod or tube. The mirrors reflect light back and forth to build up amplification.

Ruby The first laser type; a crystal of sapphire (aluminum oxide) containing trace amounts of chromium oxide.

Section The span between two SONNET network elements capable of accessing, generating and processing only SONNET section overhead; this is the lowest layer of the SONNET protocol stack with overhead.

Section overhead Nine bytes of overhead accessed, generated and processed by section terminating equipment; this overhead supports functions such as framing the signal and performance monitoring.

Semiconductor laser Lasers based on semiconductor gain media which produces its output from semiconductor materials such as GaAs.

Signal to noise ratio The ratio of the signal level to the background noise. Usually measured in decibels.

Single mode fiber An optic fiber which propagates a single mode.

Slip An overflow (deletion) or underflow (repetition) of one frame of a signal in a receiving buffer.

Solid state laser A laser where the lasing medium is a solid material such as a ruby rod. These can be optically pumped by a flash lamp or diodes. Solid state lasers also include diode lasers as they use electrically pumped solids to produce light.

Spectral width The range of wavelength emitted by a light source.

Splice A permanent means of connecting two fibers. Alternatives are fusion splice and mechanical splice.

Spot size The mathematical measurement of the radius of the laser beam.

Step index fiber A fiber in which the refractive index changes abruptly between the core and the cladding.

S t i m u l a t e d emission Emission of multi-photon of precisely the same wavelength whose wave patterns are perfectly in phase.

Stratum Level of clock source used to categories accuracy.

Superframe Any structure made of multiple frames; SONNET recognizes superframes at the DS-1 level (D4 and extended superframe) and at the VT (500 ms STS superframes).

Synchronous A network where transmission system payloads are synchronized to a master (network) clock and traced to a reference clock.

Synchronous digital hierarchy (SDH) The ITU-T-defined world standard of transmission whose base transmission level is 52Mbps (STM–0) and is equivalent to SONNET's STS-1 or OC-1 transmission rate; SDH standards were published in 1989 to address inter working between the ITU-T and ANSI transmission hierarchies.

Synchronous optical network (SONNET) A standard for optical transport that defines optical carrier levels and their electrical equivalent synchronous transport signals; SONNET allows for a multi vendor environment and positions the network for transport of new services, synchronous networking and enhanced OAM&P.

Synchronous transfer module (STM) An element of SDH transmission hierarchy; STM-1 is SDH's base level transmission rate equal to 155Mbps; higher rates of STM-4, STM-16 and STM-48 are also defined.

Synchronous transport signal level 1 (STS-1) Basic SONNET building block signal transmitted at 51.84 Mbps data rate.

Synchronous transport signal level N (STS-N) The signal obtained by multiplexing integer multiples (N) of STS-1 signals together.

Threshold The input level at which lasing begins during excitation of the laser medium.

Threshold current The lowest current that can be used to operate laser.

Total internal reflection Reflection occurring when the light approaches a change in refractive index at an angle greater than the critical angle.

Transmission Passage of electromagnetic radiation through a medium.

Transmittance The ratio of transmitted radiant energy to incident radiant energy, or the fraction of light that passes through a medium.

Vanadate lasers Lasers based on rare–earth-doped yttrium, gadolinium or lutetium vanadate crystals, usually $Nd:YVO_4$.

Vertical-cavity surface-emitting laser (VCSELs) A type of semiconductor laser diode with the optical cavity axis along the direction of current flow rather than perpendicular to the current flow as in conventional laser diodes. A low cost laser used in high-speed data communications equivalent.

Wavelength The length of the light wave, usually measured from crest to crest, which determines its color. Common units of measurement are the micrometer (micron), the nanometer, and (earlier) the Angstrom unit.

WDM (Wavelength division multiplexing) The simultaneous transmission of several signals of different wavelengths along a single fiber.

Wideband Services requiring 1.5 to 50 Mbps transport capacity.

Windows Commonly used bands of wavelengths. The first window is 850 nm, the second window is 1300 nm and the third window is 1550 nm.

YAG (Yttrium Aluminum Garnet) A widely used solid state crystal which is composed of yttrium and aluminum oxides which are doped with a small amount if the rare earth neodymium.

YAG lasers Lasers based on YAG (yttrium aluminum garnet) crystals, usually Nd:YAG.

YLF lasers Lasers based on YLF (yttrium lithium fluoride) crystals, usually Nd:YLF.

Ytterbium YAG (Yb:YAG) laser A solid state diode/flash lamp pumped laser with 1.03 μm wavelength mainly used for Optical refrigeration and materials processing.

Ytterbium-doped gain media Laser gain media containing laser-active ytterbium ions.

YVO$_4$ laser A laser created by exciting a YVO4 crystal doped with an Nd ion using an LD or lamp produces a laser beam that has the same wavelength as YAG.

Z-Cavity A term referring to the shape of the optical layout of the tubes and resonator inside a laser.

REFERENCES

1. Senior, J. M, Optical fiber communication, Prentice Hall 1992

2. Jeff Hecht, understanding fiber optics 1998

3. Byeong Ha Lee et al, 2012 Sensors, 12, 2467

4. Kim, Y.J et al, *Opt. Lett.* 2002, 27, 1297

5. Allsop, T.et al, 2002,*Rev. Sci. Instrum.* 73, 1702

6. Fu, H.Y.; Tam, H.Y.; Shao, L.Y.; Dong, X.; Wai, P.K.A.; Lu, C.; Khijwania, S.K. 2008, *Appl. Opt.*, 47, 2835

7. Dennis Hohlfeld, Ph.D Dissertation zur Erlangung des Doktorgrades der Fakultät für Angewandte Wissenschaften der Albert-Ludwigs Universität Freiburg im Breisgau, 2005

8. Siva Ram Murthy C, Guruswamy M., WDM Optical Networks, Concepts, Design, and Algorithms, Prentice Hall India, 2001

9. Ishio, H. Minowa, J. Nosu, K., "Review and status of wavelength-division-multiplexing technology and its application", 1984, Journal of Lightwave Technology, 2, 448

10. Cheung, Nim K.; Nosu Kiyoshi; Winzer, Gerhard "Guest Editorial/ Dense Wavelength Division Multiplexing Techniques for High Capacity and Multiple Access Communication Systems", IEEE Journal on Selected Areas in Communications, 1990, 8

11. Arora, A.; Subramaniam, S. Wavelength Conversion Placement in WDM Mesh Optical Networks,. Photonic Network Communications, 2002, 4

12. Frank Hiatt, Synchronous Optical net working, Technical Review, Lucent technologies, 1999

13. Wilmsen. C et al, Heterogeneous Optoectonics Integration, E. Towe, (ed). SPIE Press, Bellingham, WA, 2000.

14. Ramaswami and Sivarajan: Optical Networks :A Practical Perspective, Morgan Kaufmann Publishing, 2002.

15. Gilbert Held, Deploying Optical Networking Components, McGraw-Hill.2001

16. Govind E. Agrawal, Fiber-Optic Communications Systems, Wiley & Sons,2002

17. Osamu Wada, Optoelectronic Integrated Circuits-Chapter 27, in Handbook of Electro-Optics, (Ed. R. Waynant) McGraw-Hill, New York, 1994, 2000.

18. Higuera J.M. (Ed.) Handbook of Optical Fiber Sensing Technology, Wiley, 2004

19. C.M.Davis and C.J Zarobila, in Fiber optic sensors, Optoelectronics Materials, McGraw Hill, 1987

20. John Crisp, Barry Elliott, Introduction to fiber optics, Elsevier, 2005

21. Vincent P. Wnuk, Alexis Méndez, Steve Ferguson, Tom Graver, Process for Mounting and Packaging of Fiber Bragg Grating Strain Sensors for use in Harsh Environment Applications, *Smart Structures Conference 2005*

22. Bahareh Gholamzadeh, and Hooman Nabovati, Fiber Optic Sensors , World Academy of Science, Engineering and Technology 42, 2008

23. Tian Zhao,Yuan Gong, Yunjiang Rao,Yu Wu, Zengling Ran, and Huijuan Wu, 2011, Chinese Optics Letters 9, 050602

24. Olsson. A, Erbium Fiber Amplifiers-Fundamentals & Theory, Academic Press, 1997

25. Sudo, Optical Fiber Amplifiers: Materials, Devices, 1997, Artec. House, Boston

26. Tkach, R.W and Chraplyuy A.R, Optical and quantum electronics,(1989) 21, S105

27. Ctistis, G., Yuce, E., Hartsuiker, A., Claudon, J., Bazin, M., Gérard, J., & Vos, W. (2011). Ultimate fast optical switching of a planar microcavity in the telecom wavelength range *Applied Physics Letters, 98* (16)

28. Yano, H., G. Sasaki, M. Murata, and H. Hayashi (1992) IEEE Tran. Electron Devices, 39,. 2254

29. Hong, W.-P., G.-K, Chang, R. Bhat, J. L. Gimlett, C. K. Nguyen, G. Sasaki, and M. Koza, (1989 *Electron. Lett.*, 25, p. 1562,

30. Chandrasekhar, S., J. C. Campbell, A. G. Dentai, C. H. Joyner, G. J. Qua, A. H. Gnauck, and M. D. Feuer,(1988) *Electron. Lett.*,l. 24, 1443.

31. Lunardi, L. M., S. Chandrasekhar, A. H. Gnauck, C. A. Burrus, R. A. Hamm, 1995 *IEEE Photonics Technol. Lett.*,. 10, 1201.

32. Nagao, H., and M. Yamamoto, *IEICE Tech. Rep.*, vol. 89, p. 288, CPM89-70, 7 (1989) (in Japanese) and Sharp Electronic Components Catalog, "OPIC."

33. Joan R.Casas, Paulo J.S.Cruz, M.ASCE, 2003 Journal of bridge engineering, 8, 362.

34. F.Yu,S.Yin, *Fiber optic sensors*. Marcel-Dekker, 2002

35. *Creath.O K, and. Wyant J. C,* in Moiré and Fringe Projection Techniques optical Shop Testing, Second Edition, Edited by Daniel Malacara., John Wiley & Sons, inc.1992

36. Bosselmann, T, et al., Fiber-Optic Temperature Sensor Using Fluorescence Decay Time, *Proc. SPIE. Fiber Optic System for Mobile Piatforms*, 1991, 32

37. Farahi, F., Strain and Temperature measurement Using optical Fiber, in *Applications of Fiber Optic Sensors in Engineering Mechanies*, Farahi, F. Ed., American Society of Civil Engineers, New York, 1993, 236

38. Rao, YJ., Jackson, D.A., Joncs, R., and Shannon, C., Development of Prototupe Fiber-Optic- based Fizeau Pressure Sensors with Temperature compensation and Signal Recovery by Coherence Reading, *J. of Lightwave Tech.*, 12, 1994, 1685

39. Saaski, E., J. Hartl, and G., Mitchell, A Fiber Optic Sensing System Based on Spectral Modulation, 1986, *Adv. Instrument.*, 41, 1177, and Fiber-Optic Sensing of Physical Parameters,product feature in *Sensors*, 1988, *The Journal of Machine Perception*, . 5, . 21,

40. P´erez-Oc´on. F., Rubi˜no.M, Abril. J.M,Casanova.P , Mart´ınez.J.A, Fiber-optic liquid-level continuous gauge, 2006 Sensors and Actuators, A 125, 124

41. Peterson. J.I, and Vurek, G.G, Fiber-Optic sensors for biomedical applications, 1984 *Science*, 224, 4645, 123

42. Mignani. A.G and Baldini.F, Biomedical sensors using optical fibres, 1996 *Rep. Prog. Phys.*, 59, 1-28

43. Baldini.F , Giannetti.A, Mencaglia.A.A, and Trono.C, Fiber Optic sensors for Biomedical Applications, 2008, *Curr. Anal. Chem.*, 4, 378

44. Pinet.E, and Hamel.C, True challenges of disposable optical fiber sensors for clinical environment, Third European Workshop on Optical Fibre Sensors, EWOFS 2007, Naples, Italy.

45. Pinet.E, and Hamel.C, True challenges of disposable optical fiber sensors for clinical environment, 2007, Proc. SPIE 66191Q, 1–4

INDEX

www.ingramcontent.com/pod-product-compliance
Lightning Source LLC
Chambersburg PA
CBHW031807190326
41518CB00006B/225